Who Pays for Clean Air

Who Pays for Clean Air

The Cost and Benefit Distribution of Federal Automobile Emission Controls

David Harrison, Jr.
Harvard University

Ballinger Publishing Company • Cambridge, Mass.
A Subsidiary of J.B. Lippincott Company

This book is printed on recycled paper.

International Standard Book Number: 0-88410-451-6

Library of Congress Catalog Card Number: 75-22060

Printed in the United States of America

Library of Congress Cataloging in Publication Data

Harrison, David, 1946–
    Who pays for clean air.

    Bibliography: p.
    1. Air-pollution—Economic aspects—United States—Mathematical models.
    2. Motor vehicles—Pollution control devices—Costs. I. Title.
HC110.A4H37   629.2'53   75-22060
ISBN 0-88410-451-6

To My Mother and Father

Table of Contents

List of Figures

List of Tables

Preface

This book is an empirical study of federal automobile emissions control—a major governmental attempt to improve air quality in the United States. In the 1970 Amendments to the Clean Air Act, Congress mandated 90 percent reductions in new car emissions of carbon monoxide (CO), hydrocarbons (HC), and nitrogen oxides (NO_x) over the 1970 model year. The 1970 legislation provided that this automobile emission control program be administered by the federal Environmental Protection Agency (EPA). The EPA and other organizations have developed data on the program which are used in this book to estimate the costs and benefits of the current auto emission control program and several plausible alternatives.

This study's results were completed in May 1974 and incorporate the technological information and administrative and legislative developments through the beginning of 1974. Because of the inevitable time lag between completion of the results and the publishing date, a question naturally arises as to whether the results presented for the "current program" are still valid. I believe they are still valid and will be so until the congress decides to make substantial changes in the nature of automobile emission control.

Since this study was completed, the "current program" has been modified somewhat by two postponements of the final emission standards beyond the one year discretionary postponement permitted in the 1970 legislation and granted by the EPA in 1973. The timetable established in the 1970 amendments provided that the 90 percent reduction in new car emissions be met in the 1975 model year for CO and HC and in the 1976 model year for NO_x. The 1973 postponement resulted in a revised timetable for the "current program" to 1976 for CO and HC and 1977 for NO_x. This revised timetable is the one used in this study for the "current program." In June 1974 the final standards were postponed a further year, to 1977-78, under the Energy Supply and Environmental Coordination Act of 1974. In March 1975 the administrator of the EPA postponed the 1977 standards for CO and HC to 1978 because of some evidence that the control technology used to control these emissions may lead to increased emissions of other harmful substances.

There have also been several additional studies of the automobile emission standards published since this study was completed, the most notable one being the four volume study published by the National Academy of Sciences in September 1974. The Academy studied the technology and costs of auto emission control, the health effects of air pollutants, the meteorology of air pollution, and the overall costs and benefits of the current control program. The costs and benefits study, in which I participated, used updated technology and cost information to estimate the aggregate costs of emission control and generated estimates of the dollar value of benefits of control using several estimation techniques.

Although these recent changes somewhat alter the landscape of the auto emission control program, this study's results and conclusions are little affected for several reasons. For one thing, the new information on control costs are not substantially different from the figures used in this study, which are based largely on previous Academy estimates. Moreover, this study concentrates on *variations* in impacts for households in various groups, and these variations are not greatly affected by modest changes in the *levels* of costs and air quality benefits. Finally, the emphasis in this study is on the long run impacts of the current program, which are presented for the year 1990 in this book. The recent postponements in emission standards may mean that the estimates in this book are more appropriate for 1992 or 1993 rather than 1990, but that would be a minor change.

Other natural preliminary questions are what readership the book addresses and what technical background is assumed. This book is designed to be of interest to a wide variety of readers interested in analysis of major public policies. While it is an economic study, every effort has been made to purge the book of confusing jargon and to segregate technical matter in specific chapters and appendixes. There is a considerable amount of technical material in Part I of the book, where the procedures and data are described. This technical material is necessary to document the procedures and the data used to generate the empirical results. However, the first chapter of Part I provides an overview of the procedures. Some readers may wish to skim the remainder of Part I and concentrate on the results and conclusions which are given in Part II of the book.

I have accumulated many debts of gratitude in the course of this three year enterprise. During the first two years, the project was my Ph.D. dissertation in Economics, submitted to Harvard University in September 1974. In the last academic year, 1974–75, the material was substantially rewritten for book form. The number of people who helped me in one way or another during these years is very large, and not all can be mentioned although all are appreciated. Of course I alone am responsible for any errors or omissions in the final product.

My major intellectual debt is to Professor John Kain, chairman of my dissertation committee and my principal mentor for several years. His advice in the formative stages of the study was particularly helpful, although his help extended to all aspects of the undertaking. I am deeply grateful for this assistance. I was fortunate to have two other faculty members on my dissertation committee at different times. Both Professor Richard Caves and Professor John Meyer were very helpful and encouraging, and I would like to thank them both. In addition, I am grateful for insightful comments I received from Professor Gregory Ingram at several points in the study. My colleague in the City Planning Department, Professor Gary Fauth, was of great help in the later revisions of the book, and his perceptive remarks notably improved the final product.

I have received financial support over these years from several sources. As a graduate student, I was the recipient of the Graduate Prize Fellowship, funded by the Ford Foundation. The beginning of the study was formulated during a summer at the Urban Institute, and I thank Dr. Harold Hochman for his support. Part of the study was written while I was an employee at the Transportation Systems Center of the U.S. Department of Transportation. I would particularly like to thank Dr. Frank Hassler and Dr. William Duffey for letting me pursue these matters. The National Science Foundation also provided support through a grant for me and several others to study the automobile and the regulation of its impact on the environment.

Several persons aided in the large tasks of data gathering and computer programming. Craig Schweinhart performed most of the programming, and Richard Shepro provided research assistance and some programming help. Robert MacDonald helped to edit several drafts of the manuscript, and his discerning remarks went far beyond more editorial help to greatly enhance both the logic and the readability of the book. Jenifer Lyons was responsible for a superb typing job on the final draft. Jeanne Dernback and Madeleine Lane very ably typed earlier drafts.

Finally, I would like to thank my wife, Alexandra Murray Harrison, both for editorial help at several points in the study and for more general encouragement throughout.

David Harrison, Jr.
Cambridge, Massachusetts
June 1975

Chapter One

Introduction

Americans have become increasingly aware of their physical environment—the air they breathe, the water they drink, the noise they hear, and the land they use. This awareness is reflected in a growing tendency to notice the environmental impacts of various activities and to take steps to minimize the ill effects. A spate of environmental laws has been passed at all levels of government in recent years, but the federal government has taken the leadership in several areas of major environmental concern. Air pollution control is one of these areas.

THE 1970 CLEAN AIR ACT AMENDMENTS AND AUTOMOBILE EMISSION CONTROLS

Major federal initiative in air pollution control was taken in the 1970 Amendments to the Clean Air Act of 1963.[1] Before 1970 the federal government primarily played an advisory role, leaving the setting of air quality goals and the implementation and enforcement of air quality improvement programs largely to the states. Relatively little state action was taken except in California. During the 1960's the state of California embarked on a serious effort to improve air quality in its large urban areas, particularly in the Los Angeles air basin. These efforts concentrated on controlling emissions from automobiles since, as a result of work done in the 1950's by Dr. Haagen-Smit and others at the California Institute of Technology, automobile emissions were identified as being largely responsible for the steadily worsening air quality in Southern California.

Dr. Haagen-Smit determined that the eye-irritating portion of the brown haze or "smog" that engulfed Los Angeles consisted of photochemical oxidants (O_x), which were not emitted directly into the atmosphere but were instead produced by complicated atmospheric reactions involving nitrogen oxides (NO_x), hydrocarbons (HC), and sunlight. Although both NO_x and HC are

1

emitted from stationary sources as well, automobile emissions were found to be the major source of both these pollutants in the Los Angeles basin. Thus the blame for the watery eyes, stuffed-up noses, and hazy skies that Los Angeles residents suffered was placed squarely on the automobile.

These California studies inspired other studies on automobile pollution and its adverse effects. Air pollution damage research indicated potentially troubling effects of O_x concentrations on human health, particularly respiratory diseases, and on vegetation and materials. Studies also indicated that NO_x concentrations have independent damaging effects apart from their contribution to concentrations of photochemical oxidants. In addition, automobile emissions were determined to be the dominant source of another air pollutant, carbon monoxide (CO), which was linked to human health effects. Monitoring data in other cities in California and elsewhere indicated that automobile pollution was not peculiar to Los Angeles; these auto pollutants were present in many other areas at levels far exceeding the background or natural levels.

The state of California began to require emission controls on autos sold in California in the 1961 model year, when a simple crankcase control device was required to control HC emissions. Modest exhaust emission controls on HC and CO went into effect in the 1966 model year, after technology to control exhaust emissions was determined to be effective. After initial protests about the validity of the link between auto emissions and air pollution and the feasibility of emission control, the automobile companies complied with the California standards for new car emissions. There were no great objections when the emission standards for 1966 California cars were extended by the federal government to cover all cars sold in the United States in the 1968 model year.

That peace was shattered by the 1970 amendments. The 1970 legislation mandated a schedule to reduce automobile emissions of CO, HC, and NO_x which entailed a 90 percent reduction from the 1970 controlled levels. The 90 percent reduction was to be achieved by the 1975 model year for CO and HC and by the 1976 model year for NO_x. Since the legislation mandated percentum reductions, it was left to the United States Environmental Protection Agency (EPA) to set specific emissions standards in grams per mile (g/mi) and resolve the specifics of test procedures, compliance procedures, and the like. Administratively, the final standards were set at 3.4 g/mi for CO, 0.41 g/mi for HC, and 0.4 g/mi for NO_x.[2]

Since 1970 the auto industry has devised various emission control systems while lobbying for less stringent regulations. When the 1970 amendments were passed, the automobile companies did not have the technology to provide 90 percent reductions from 1970 levels, and many companies publicly proclaimed that such technology could not be developed under the severe time constraints imposed by the law. The 1970 amendments provided for one year extensions of the deadline for full reduction in new car emissions of HC and CO to 1976 and of NO_x to 1977. These extensions were contingent on the automobile companies showing that the technology to reduce emissions to their final levels would not be available by the earlier date and that they had made a good faith effort to develop the necessary control technology. After considerable public debate and legal action, in 1973 the automobile companies won all three extensions.[3]

AIM OF THIS ECONOMIC STUDY

How will the blessings and burdens of this federal auto emission control program be distributed among groups in society? How will different strata of society—different "income groups"—absorb the costs and share the benefits of such a program? Will urban and rural households be affected in the same way? Will some regions of the country be favored or harmed more than others? How will these different groups fare under alternative control strategies? This study answers these questions by estimating the distributional patterns of benefits and costs of automobile emission control for household in various income groups and various geographical areas in the United States.

These questions are different than those usually addressed in economic studies of public policy. Economists have traditionally been concerned with the aggregate costs and benefits of government programs and seldom with the distribution of costs and benefits among households in various income groups and geographic areas. There seem to be two major reasons for economists' reluctance to consider distributional effects, one theoretical and the other practical.

The theoretical reason stems from the concern for economic efficiency. Economic efficiency relates to the overall size of the national income. Economists have customarily been foes of taxes and other "distortions" which serve to reduce national income, even when these taxes or distortions tend to benefit disadvantaged groups. According to this view, the economist should not impose his value judgments about equity effects but rather should provide a value neutral judgment about the overall impacts of the scheme on the economy. The concern for efficiency carries over into evaluations of public sector projects or regulatory programs. Evaluating the efficiency of public sector programs involves determining whether the total benefits of a proposed project or policy exceed its costs. If the benefits do exceed the costs, the project will be a net contribution to national welfare and, using the economic efficiency criterion, should be undertaken. According to this calculus, it is not necessary to know the distribution of a project's costs and benefits among households. If the society wants to redistribute income, it is argued, direct payments should be made.

The practical reason for ignoring distributional effects is that it is more difficult to estimate the separate effects of government programs on households in different income groups and geographic areas than it is to estimate total costs and benefits. The data needed to determine the income profile of the households paying the costs or receiving the benefits of government programs are not readily available. Data on the geographic distribution of costs and benefits are no easier to obtain.

However, government policies significantly affect the distribution of income in the society, and an analysis which ignores these considerations is of limited usefulness. Distributional considerations should be estimated and presented as economic implications along with the aggregate costs and benefits. While it is true that the distributional impacts are more difficult to estimate, modern computer facilities make calculation of quite disaggregated effects a manageable task.

The focus on distributional effects is not at the expense of analysis of the consequences for economic efficiency. This study aggregates the separate cost and benefit estimates to obtain estimates of the total costs and total benefits of the current automobile emission control scheme and several plausible alternatives. These figures are used to assess the cost-effectiveness of the alternatives. However, this is not a true cost-benefit analysis because the benefits of improved air quality are not given a dollar value. Benefits are measured as reductions in concentrations of air pollutants.

PLAN OF THE STUDY

This study is divided into two major parts. Part I describes and documents the estimation procedures while Part II presents the results and conclusions of the study. The procedures in Part I are discussed in terms of the current program, although the same framework is used to assess the impacts of alternative schemes. Chapter Two is an overview of the procedures and the data used to estimate cost burdens and benefits received by various income and geographic groups. The technological, cost, and pricing information needed to estimate emission control costs for the current program are derived in Chapter Three. The next two chapters, Chapters Four and Five, provide detailed derivations of the emission control cost burdens and air quality benefits for household groups.

The first two chapters of Part II constitute the heart of the empirical analysis of the current auto emission control strategy. Chapter Six deals with the costs and Chapter Seven deals with the benefits of the scheme. Both chapters provide national estimates and estimates of variations by income group, urban area, and region of the country. The results for the current program provide the motivation for comparisons of the current scheme with several plausible alternative automobile emission control strategies. Chapter Eight compares these strategies in terms of their distributional impacts and their cost-effectiveness. Chapter Nine is a summary of the major conclusions of the study.

NOTES TO CHAPTER ONE

1. Public Law No. 91–604 § 4 (a), 84 Stat. 1676 (1970).
2. 40 C.F.R. § 85.075-1 (a) (1973); 40 C.F.R. § 85.21 (b) (3) (1972).
3. The emission standards for CO and HC originally promulgated by EPA for 1975 were postponed by the administrator to 1976 on April 11, 1973. Environmental Protection Agency, Motor Vehicle Pollution Control Suspension Granted, Decision of the Administrator, April 11, 1973, 38 *Fed. Ref.* 10317 (1973). The 1976 standard for NO_x was postponed to 1977 on July 30, 1973. *In re* Applications for Suspension of 1976 Motor Vehicle Exhaust Standards, Decision of the Environmental Protection Agency Administrator (July 30, 1973), recorded in *Hearings Before the Subcommittee on Air and Water Pollution of the Senate Committee on Public Works on the Decision of the Administrator of the Environmental Protection Agency Regarding Suspension of the 1975 Auto Emission Standards,* 93rd Cong., 2d sess. 2083, 2086–87 (1973). This study's evaluation of the current automobile emission control program includes these extensions of the final standards to 1976–77.

Part I

The Estimation Procedures

Chapter Two

Overview of the Estimation Procedures and the Data

The major purpose of this chapter is to overview the procedures used to estimate the impacts of auto emission control on households in various income groups and geographic areas. The cost impact procedures and benefit impact procedures are discussed separately. This separation simplifies the explanation, although there are considerable common elements in the two procedures. The division also corresponds to the presentation of results for the current program in Part II, which includes a chapter on cost impacts and a chapter on benefit impacts. The overviews are preceded by a discussion of the relation of these procedures to those of other studies that have estimated the economic impacts of auto emission control.

This chapter has several other purposes, served by the sections coming after the overviews. The section after the overviews discusses the important distinction between absolute and relative costs and benefits. Relative cost and benefit calculations are used extensively in Part II to evaluate the income distributional consequences of auto emission controls. The data on automobile ownership and travel behavior used in this study are then described. This data was crucial; the cost and benefit estimates derived in this study would not have been possible without very detailed data on car ownership and travel. The chapter concludes with a discussion of the limitations of the estimation procedures.

RELATION TO OTHER STUDIES

Several recent studies have evaluated the automobile emission control strategy undertaken under the 1970 Amendments to the Clean Air Act. These studies fall into two basic categories: estimates of national costs and benefits, and detailed estimates for one urban area. Most are of the first type. These national estimates do not break down the costs and benefits by geographic area or income group, although they often present figures for several years. The second estimation methodology is more complicated. Authors of such detailed studies develop models to simulate the meteorology of an urban area and generate estimates of air quality improvements for many subareas within the urban area.

These two approaches have complementary strengths and weaknesses. The national estimates are simple and comprehensive. Costs and benefits accruing to all Americans are obtained and total costs and benefits can be compared over time. The 1973 report of the administrator of the Environmental Protection Agency to the Congress, "The Cost of Clean Air,"[1] is an example of a national study. The report estimates the costs and benefits of the (then) 1975-76 automobile emission standards. Benefits are measured by reductions in total emissions. The national CO, HC, and NO_x emissions from automobiles in 1978 with and without the 1975-76 standards are given as:

CO: 52 million tons with control or 74 million tons without

HC: 6 million tons with control or 8 million tons without

NO_x: 3.1 million tons with control or 4.6 million tons without

The report estimated that the annual costs of these controls would be approximately $6 billion (1970 dollars) in 1978. These costs include the annual investment costs for emission control equipment and the increased annual maintenance and fuel costs due to emission control equipment.

The costs and benefits estimated by other national studies differ somewhat, although they use essentially the same methodology.[2] To obtain total national costs, increases in per car manufacturing costs, annual maintenance costs, and annual fuel costs are estimated and then multiplied by the total number of cars sold. Total national emissions benefits are obtained by estimating emissions characteristics of cars with and without controls, aggregating to obtain total emissions with and without controls, and calculating the difference.

The national studies provide an overview of the impact of the emission control strategy. But much useful detail is ignored. The costs and benefits accruing to particular household groups cannot be determined. In addition, these studies usually measure benefits as tons of pollutants prevented. Reducing tons of pollutant emission will reduce households' exposure to high air pollutant concentrations. But a more direct measure is preferable.

Detailed urban air pollution models provide better estimates of air quality benefits since they directly estimate air pollutant concentration improvements from auto emission controls. For instance, the Transportation and Air

Shed Simulation Model (TASSIM), developed recently by Gregory Ingram and others, estimates reductions in air pollutant concentrations for 122 subareas in the Boston region.[3] The TASSIM model was designed to evaluate the air quality impacts of federal auto emission controls as well as a host of more localized control schemes, such as restrictions on automobile access to downtown areas or improvements in public transportation.

The virtues of the urban modeling procedure for assessing the impacts of federal auto emission controls are won at the expense of completeness and simplicity. Since these models are constructed for specific urban areas, the results are not easily extrapolated to other urban areas. In addition, the precision of the benefit analysis is not matched by a correspondingly detailed analysis of the costs. Costs are calculated in the same way as in national estimates, and the costs accruing to households in different income groups are no easier to estimate in a spatially detailed urban model than in a national model. Since these models are complex, it is difficult to judge the accuracy of their estimates or the validity of their underlying assumptions.

The estimation procedure in this study lies in the middle ground between these two procedures. This study's procedure produces estimates of the costs and benefits accruing to household groups in the United States, in contrast to the national models which provide only totals; but these household groups are not as detailed as the groupings possible using detailed meteorological models of one urban area.

Specifically, this study generates cost and benefit estimates for 3,409 subgroups in the United States. Figure 2-1 depicts the nature of these

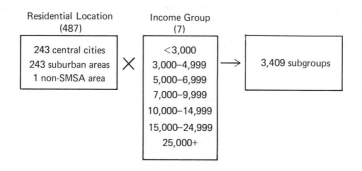

Figure 2-1. Household Subgroups in the Study

household subgroups. Households are crossclassified by their income group and their place of residence. A total of 487 geographic areas are identified, representing the central city and suburban portions of the 243 Standard Metropolitan Statistical Areas (SMSAs) identified by the census and one residual non-SMSA area. Within each geographic area, households are categorized into seven income groups, ranging from less than \$3,000 per year to more than \$25,000 per year. The product of the seven income groups and 487 geographic areas gives the 3,409 subgroups identified in this study. However, the actual number of costs and benefit estimates is greater than 3,409 because estimates are prepared for two years, 1980 and 1990, and because air quality benefits are measured by improvements in concentrations of three air pollutants.

Before overviewing the cost and benefit estimation procedures, some comment on the notation used in the overviews would be useful. After describing the estimation procedures in words, the steps are summarized using subscripted mathematical notation. Following convention, the subscripts denote that there are several estimates for each category. For example, O is average ownership costs and the subscript i refers to income group. So O_i refers to seven estimates of annual ownership costs, one for each of the seven household income groups. Many of the procedures involve summing terms. This summation process is shown by the symbol Σ, given with indications of the category being summed and the range of summation. For instance, using N to denote the number of households, the product $O_1 \cdot N_1$ is the ownership costs borne by all households in the first (or lowest) income group. The term $\sum_{i=1}^{7} O_i N_i$ refers to the sum,

$$(O_1 \cdot N_1) + (O_2 \cdot N_2) + (O_3 \cdot N_3) + (O_4 \cdot N_4) + (O_5 \cdot N_5) + (O_6 \cdot N_6) + (O_7 \cdot N_7),$$ or the total ownership costs borne by all income groups.

This mathematical notation would be needlessly complicated if all four major categories—household income group, geographic area, analysis year, and air pollutant—were included. There is no reason to differentiate the geographic area and air pollutant in the overview since the procedures are the same for all areas and air pollutants. Thus, the overviews in the following sections give procedures to estimate the costs and the benefits accruing to various income groups in each analysis year. It is understood that the same procedures apply to all 487 geographic areas and all three air pollutants.

COSTS OF AUTOMOBILE EMISSION CONTROL

Households as a group will ultimately bear all the costs of automobile emission control. The first task is to identify the major ways the burden is borne. Many of the costs of automobile emission control are initially incurred by automobile companies in their efforts to develop and install control equipment. But most or all of these auto company costs are eventually reflected in new car prices, and these new car price increases raise the cost of automobile ownership. So the first way that households bear emission control costs is as car owners paying more for their cars. Car owners also incur greater operating costs for fuel and maintenance because emission controls result in worsened fuel economy and increased repair and maintenance expenses.

If part of the automobile company expenses for emission control are not passed on to car buyers, automobile company stockholders will bear some of the costs in the form of reduced corporate profits. Since these reduced corporate profits result in decreased federal corporate income tax receipts, federal taxpayers will also bear some of the costs of emission control.

These, then, are the four categories of cost bearing identified in the estimation procedure: (1) increased automobile ownership costs, (2) increased automobile operating costs, (3) reduced automobile company stockholder profits, and (4) reduced federal tax receipts.[a] This study develops procedures to estimate the average costs borne by each income group in each of these four categories.

Automobile Ownership Costs

Consumers pay more for automobile ownership when emission control costs are passed on to car purchasers. In the first instance these increased prices are borne by new car purchasers. As cars are sold in used car markets, part of the increased new car purchase price is passed on to used car buyers.

To estimate the average ownership costs borne by households in particular income groups in 1980 and 1990, this study relies on automobile ownership data rather than on automobile purchase data. The car ownership cost estimates in this study are consistent with a user cost view of automobile ownership, in which automobile ownership is equivalent to rental of a capital good.[4] The annual rental charge includes a payment of depreciation and an interest payment for the use of capital. In this framework, increased car prices raise the price of automobile capital and thereby increase annual depreciation and finance charges.

Depreciation costs are not directly paid each year but are implicit in the automobile's purchase price. Like any durable good, the automobile deteriorates over time as parts wear out and maintenance costs increase. This physical deterioration is reflected in a declining price of the car as it ages. The change in the price of the car from one year to the next measures the implicit depreciation costs paid by the car owner. Finance costs are often more explicit since many car owners finance the purchase of their car.

An example may be useful. In 1990 every car owner will bear some increased depreciation and finance costs because of the current auto emission control strategy. Consider an owner of a new (1990 model) car who pays, say, an additional $300 for the car because of emission controls. The finance costs

---

[a]Several other ways that households may bear some of the costs of automobile emission control are not included in this four part list. For example, automobile dealers may bear some of the costs if their profit margins are reduced by cost and price adjustments. Some of the control costs may be borne by automobile company workers in the form of lower wages. These and other possible adjustments are not included in the analysis partly because they are quite speculative and partly because these impacts will certainly be dwarfed by the costs borne by households under the four categories identified.

he pays during 1990 will include the interest on this $300. The implicit depreciation charge will also increase. If the owner could recoup $200 of the $300 increase upon selling the car at the end of 1990, the increase in depreciation due to emission controls would be $100. The same logic would apply to owners of 1989 cars, 1988 cars, and so on.

Ownership Costs by Car Age. The formula for calculating the increased annual ownership costs for a car in age group j in year t is given by the following equation:

$$O_j^t = (P_j^t - P_j^{t+1}) + (r \cdot P_j^t) \tag{2.1}$$

where: O = ownership cost burden

   P = price increase due to emission controls

   r = interest rate

   j = car age group

   t = analysis year

The left-hand and right-hand parentheses measure the increased depreciation charge and the increased finance charge, respectively. In Chapter Three car price increases are estimated so that annual increases in car ownership for six car age groups can be calculated for 1980 and 1990.

Ownership Costs by Income Group. To calculate the average ownership cost borne by households in a particular income group, the car age costs are weighted by the average number of cars in each car age group owned by members of that income group. For ease of exposition, these car ownership averages are termed *probabilities of car ownership.*[b] The formula for the average ownership cost burden borne by a household in the ith income group is the following:

$$O_i^t = \sum_{j=1}^{6} (s_{ij}^t \cdot O_j^t) \tag{2.2}$$

where: O = ownership cost burden

   s = probability of car ownership

   i = income group

   j = car age group

   t = analysis year

---

[b]The average car ownership by car age data are not, technically speaking, probabilities since they do not add up to 1.0 when summed over all car age categories. The derivation of the S_{ij}S in Equation 2.2 is explained later in this chapter.

Automobile Operating Costs

All owners of controlled cars will also incur greater annual operating costs because emission control modifications diminish fuel economy and necessitate additional repair and maintenance costs. Although some car owners may not be aware of the size of these penalties, these costs of emission control will be reflected in greater gasoline expenses and greater payments to garages for maintenance and repair.

Operating Costs by Car Age. Fuel and maintenance penalties vary with emission control technology, and thus with model year, just as car price penalties do. Maintenance penalties for cars of various ages are measured as *annual* dollar cost increases, denoted as M_j in the calculations. Fuel penalties are expressed as increases in *per mile* fuel costs, calculated from estimates of reductions in gas mileage (miles per gallon) and the price of gasoline. These fuel penalties by car age group are represented as a_j in the estimation procedure. The M_js and the a_js for the six car age groups in 1980 and 1990 are estimated in Chapter Three.

Operating Costs by Income Group. A household's annual increase in fuel expenses due to emission controls depends on the number of miles it travels during the year and, because the fuel penalties differ by model year, on the vintage of car driven. To obtain average fuel increases for households in each income group, estimates of the annual miles driven are used as weights along with the per mile penalties. The mileage estimates for each income group are the average miles traveled in each car age group by households in that income group.

The other component of operating costs, increased annual repair and maintenance costs, is calculated using the same car ownership probabilities and the same procedure used to calculate ownership cost. The total annual operating cost increase for a household in the ith income group is the sum of additional fuel costs and additional repair and maintenance costs, as summarized in Equation 2.3.

$$U_i^t = \sum_{j=1}^{6} (a_j^t \cdot V_{ij}^t) + \sum_{j=1}^{6} (s_{ij}^t \cdot M_j^t) \qquad (2.3)$$

where: U = operating cost burden

a = per mile fuel cost increase

V = vehicle miles of travel

s = probability of car ownership

M = maintenance cost increase

$$i \ = \ \text{income group}$$

$$j \ = \ \text{car age group}$$

$$t \ = \ \text{analysis year}$$

Stockholder Costs and Taxpayer Costs

The last two cost burden categories, stockholder costs and taxpayer costs, both depend upon the impact of emission controls on automobile company profits in 1980 and 1990. Pretax automobile company profits could either increase or decrease depending upon whether the new car price rise is greater than or less than the per car cost of emission controls. If profits fall because of emission control costs, automobile company stockholders will obtain fewer dividends or retained earnings. Taxpayers will share this loss as corporate taxes will decline.

Chapter Four presents estimates of the decrease in pretax profits due to emission controls. The pretax profit change is divided evenly between stockholders and federal taxpayers because the corporate profits tax is approximately 50 percent.

Stockholder Costs by Income Group. The decline in stockholder posttax profits is allocated to households in various income groups on the basis of their share in 1970 total corporate dividends. This formula is given below as Equation 2.4.

$$S_i = \frac{d_i \cdot .5\,A}{F_i} \qquad\qquad (2.4)$$

where: $\quad S \ = \ $ average stockholder cost burden

$\qquad\quad d \ = \ $ fraction of total dividends received

$\qquad\quad A \ = \ $ pretax decrease in automobile company profits

$\qquad\quad F \ = \ $ number of households

$\qquad\quad i \ = \ $ income group

These estimates are the same for all geographic areas, since one assumes that households in the same income group have the same likelihood of being stockholders regardless of where they live.

Taxpayer Costs by Income Group. The same procedure is used to allocate the decline in tax receipts to households in different income groups. The decline in federal corporate profits taxes is assumed to be made up for by a proportional increase in all federal taxes so that the incidence pattern is the same as for all federal taxes in 1970. Equation 2.5 gives the formula for the taxpayer burden of a household in income group *i.*

$$T_i = \frac{f_i \cdot .5A}{F_i}$$ (2.5)

where: T =   average taxpayer burden

f =   fraction of total federal taxes paid

A =   pretax decrease in automobile company profits

F =   number of households

i =   income group

BENEFITS OF AUTOMOBILE EMISSION CONTROL

Reduction in emissions brought about by the federal automobile emission control strategy promise to improve the air quality experienced by American households in all groups. Pollutants emitted by autos now contribute importantly to high urban concentrations of three air pollutants—CO, NO_x, and O_x. Were auto emissions to go unchecked, the 1980 and 1990 concentrations of these pollutants would likely be at least as high. While concentrations of other important air pollutants are not appreciably reduced by the automobile emission program, there is no doubt that air quality will be substantially improved in American urban areas because of the auto emission program.

The air quality improvements will in turn lead to a wide variety of benefits, including health improvements, reductions in materials and plant damages, and aesthetic improvements. There is, however, considerable uncertainty about the precise nature and particularly the precise magnitudes of these benefits. More problems arise in trying to place a dollar value on these benefits from better air quality. In addition, it is likely that different persons will place different values on these benefits. For example, those who place a high value on the aesthetic qualities of urban air will certainly benefit more from a given air quality improvement than those who are relatively unconcerned about these aesthetic considerations.

This study does not place a dollar value on air quality benefits. Instead, benefits are measured as reductions in the concentrations of the three pollutants, CO, NO_x, and O_x. This lack of a dollar value for benefits means that no calculation of benefits minus costs is possible. Dollar values for air quality benefits were not estimated because it was thought that the current state of knowledge on the ill effects of these air pollutants is too limited to provide reasonable values, particularly since there may be systematic differences in the values households in different income groups and different geographic areas place on air quality improvements. Appendix A of this study describes both the problems in estimating dollar values for air quality improvements and some of the efforts made to estimate the dollar value of aggregate air quality improvements. This appendix also presents a systematic review of the existing literature on the adverse effects of the three auto pollutants.

The task in benefit estimation, then, is to estimate the improvement in air pollutant concentrations received by households in each geographic area and each income group. The baseline for the benefit calculation is the concentration which would prevail in the area in either 1980 or 1990 had the 1970 amendments not been passed. The benefit estimation procedures are broken down into the following four steps:

1. determine car emission rates (emissions per mile)
2. calculate the total automobile emissions in the area
3. estimate urban area concentrations
4. determine average air quality benefits for each geographic area and using these figures, for each group within an urban area

Emissions Per Mile

The first step in estimating the benefits of auto emission control in 1980 and 1990 is to determine the average emission rates for cars on the road in those years. Although predicting average emissions for, say, a 1978 car in 1980 can be a complicated chore involving many factors and judgments, the Environmental Protection Agency has identified three major factors in the calculation of average emission rates: original emission rates for the model year car, deterioration factors for the control mechanisms over time, and factors relating emission rates to average speed.[5] The EPA formula for calculating average emissions from a car in age group j in year t is the following:

$$e_{jt} = c_j \cdot d_{jt} \cdot s_j \tag{2.6}$$

where: e = exhaust emissions per mile

c = original new car emission rate

d = deterioration factor

s = speed correction factor

j = car age group

t = analysis year

Emissions per mile by car age in 1980 and 1990 are calculated for both the uncontrolled case, in which all model year cars are assumed to meet the

1970 standards, and the controlled case in which all model year cars meet the requirements of the current emission control program. The differences between the controlled and uncontrolled emissions rates are the starting points for the calculation of the air quality benefits of the control program.

Urban Area Automobile Emissions

The next step is to combine these per mile emission rates with estimates of the number of miles driven by cars of various vintages in 1980 and 1990. Total urban area automobile emissions are calculated by multiplying the per mile emissions by the mileage figures and summing over the car age categories, i.e., Equation 2.7.

$$E_t = \sum_{j=1}^{6} (VMT_{jt} \cdot e_{jt}) \tag{2.7}$$

where: E =   total automobile emissions

VMT =   total vehicle miles of travel

e =   emissions per mile

j =   car age group

t =   analysis year

Urban Area Air Pollutant Concentration

Pollutant concentrations are predicted in this model by a technique known as proportional rollback.[6] The rollback technique assumes that pollutant concentration is proportional to total emissions. This assumption is convenient since all that is needed are estimates of the concentration and emissions of each pollutant in a base year, 1970 in this study, and emissions in the analysis year to predict concentrations in the analysis year. The algebraic expression of the rollback method used in this study is given in Equation 2.8.

$$C_t = \left[\frac{E_t + (r \cdot E_b)}{E_b + (r \cdot E_b)} \right] C_b \tag{2.8}$$

where: C =   concentration

E =   total automobile emissions

r =   nonautomobile emissions/automobile emissions

t =   analysis year

b =   base year

This rollback formulation rests on an assumption of stability: non-automobile emissions, the spatial and temporal distribution of emissions, and meteorological factors are the same in the base year and the analysis year.

Air Quality Benefits

The air quality benefit for a household in a given geographic area equals the difference in pollutant concentration with and without the control strategy, i.e., Equation 2.9.

$$B_t = C_t^u - C_t^c \qquad (2.9)$$

where: B = air quality benefit

C^u = uncontrolled concentration (i.e., *without* the 1970 Clean Air Act auto emission standards)

C^c = controlled concentration (i.e., *with* the 1970 Clean Air Act auto emission standards)

t = analysis year

Average benefits for households in the ith income group in each SMSA are a weighted average of the benefits predicted for the central city and suburban portions, i.e., Equation 2.10.

$$B_i = (B^c \cdot f_i^c) + (B^s \cdot f_i^s) \qquad (2.10)$$

where: B = average air quality benefit

f = fraction of households

i = income group

c = central city

s = suburb

$(f_i^c + f_i^s = 1)$

RELATIVE COSTS AND BENEFITS

The foregoing procedures generate estimates of dollar costs and air quality benefits. These raw figures are used to calculate total costs and benefits of auto emission control. But comparing these costs and benefits across income groups is misleading. A $50 per year cost will mean much more to a family earning $3,000 than it will to one earning $25,000. This section defines *relative* costs and *relative* benefits and uses these definitions to develop summary measures of the income distributional consequences of emission controls. These summary measures are used extensively in Part II of this study.

Relative Costs

Comparing burdens borne by households in different income groups is common in economic analyses of taxes, and widely accepted procedures and terminologies have emerged from these tax studies. Dollar burdens are divided by income to obtain a measure of the relative burden for each income group. Multiplied by 100, this ratio measures the *percentage* of income taken by the tax for households in each income group. It is these percentage figures which are compared across income groups to evaluate the income distributional impacts (or incidence) of the tax. The tax is said to be *regressive* if the percentage of income represented by the tax payment is greater for households in lower income groups. The tax is *progressive* if the percentage is greater for households in higher income groups. If the percentage is the same for all income groups, the tax is said to be *proportional.*

This study uses the same procedure and terminology to compare the burdens of emission control costs among households in different income groups. The relative burden of emission controls for a household in the ith income group is equal to the dollar cost burden divided by average income, multiplied by 100 to give a percentage figure, i.e., Equation 2.11.

$$R_i = \frac{C_i}{Y_i} \cdot 100 \tag{2.11}$$

where: R = relative cost burden

C = dollar cost burden

Y = average income

i = income group

Inequality Ratio

Relative costs are presented in this study for seven income groups, and the full income distributional implications relate to this entire pattern of cost bearing. But it is useful to have a summary measure of the cost burden pattern so that patterns in different geographic areas or different time periods can be compared. This study defines an inequality ratio to measure the degree of regressivity of the cost burden pattern.

The inequality ratio equals the relative burden for the lowest income group divided by the relative burden for the highest income group (Equation 2.12).

$$I = \frac{R_1}{R_7} \tag{2.12}$$

where: I =   inequality ratio

R_1 =   relative cost burden of a household in the lowest income group ($< \$3,000$)

R_7 =   relative cost burden of a household in the highest income group ($\$25,000+$)

Naturally, the higher this inequality ratio, the more regressive is the distribution pattern.

Relative Benefits

Comparing benefits received by different income groups is more difficult than comparing costs because benefits are expressed in improvements in air pollutant concentrations rather than in dollars. Ideally these physical benefits would be translated into dollar values and then relative benefits would be calculated in the same way as relative costs. But, as discussed in the overview of the benefit estimation procedure, it is not now possible to determine the dollar values that households in various income groups and various geographic areas place on air quality benefits. Thus, some proxy for relative benefits, that is, benefits as a percentage of income, has to be developed.

Two possible proxies for relative benefits were considered in this study. They embody different assumptions about the relationship between households' valuations of air quality benefits and their income group.[c] The first proxy is simply the physical air quality benefits themselves. This proxy assumes that average valuation of physical improvements is proportional to household income, that is, that higher income groups place a greater dollar value on given air quality benefits. The second proxy is the physical concentration benefits divided by average income, i.e., Equation 2.13.

$$R_i = \frac{B_i}{Y_i} \tag{2.13}$$

where: R =   relative benefit

B =   average air quality benefits

Y =   average income

i =   income group

[c]Appendix A discusses the possibility of differences in dollar values by income group.

This formulation assumes that households in different income groups place the same dollar value on a given improvement in air quality.

Pro-poor Ratio

As with the costs, benefits are calculated for seven income groups, and the complete distributional results relate to all seven income groups. But, as with costs, it is useful to have a summary measure of the distribution of benefits among income groups so that patterns can be compared between geographic areas and time periods. Procedures and terminology for assessing differences in benefit impact patterns are not nearly as developed as those used to assess cost differences via the tax incidence methodology. This study defines a pro-poor ratio to measure the degree of pro-poorness of benefits. This ratio is a natural analog to the inequality ratio calculated to summarize the cost burden pattern.

The pro-poor ratio is defined as the ratio of the relative benefits received by the lowest income group to the relative benefits received by the highest income group. Since two proxies were identified for relative benefits, two pro-poor ratios can be calculated. The measure using physical benefits is given below as Equation 2.14 and the measure using physical benefits divided by average income is given below as Equation 2.15.

$$P = \frac{B_1}{B_7} \tag{2.14}$$

$$P = \frac{(B_1/Y_1)}{(B_7/Y_7)} \tag{2.15}$$

where: P = pro-poor ratio

B_1 = physical air quality benefit received by a household in the lowest income group ($< \$3,000$)

B_7 = physical air quality benefit received by a household in the highest income group ($\$25,000+$)

Y_1 = average income for a household in the lowest income group

Y_7 = average income for a household in the highest income group

Equation 2.15 of course results in a much more pro-poor estimate of benefits than the first ratio.

The results presented in Part II of this study use the formulation in Equation 2.14 to calculate the pro-poor ratio. Some results using the other formulation are provided in footnotes. The fomulation of Equation 2.14 may understate the pro-poorness of benefits if households' valuation of air quality benefits does not vary by income group as much as implicitly assumed in that measure. But notice that the choice of a pro-poor ratio formula does not bias *comparisons* among different geographic areas unless there are systematic differences in the relative valuations of high and low income groups between geographic areas. Since there is no strong a priori reason to believe that systematic differences of this kind exist, the comparisons of income distributional patterns discussed in Chapter Nine are not affected by selecting one or the other pro-poor formulation.

DATA ON AUTOMOBILE OWNERSHIP AND TRAVEL

Car ownership and travel data are vital elements in the estimation of both the costs and the benefits of auto emission control. The cost borne by a household depends primarily on the number and age distribution of cars it owns and on the miles the household drives. The benefits depend on car ownership and travel behavior in the overall geographic area, since each household's benefits are only negligibly influenced by its own car ownership and travel characteristics. This section describes these crucial car ownership and travel data.

Car Ownership by Car Age and Income Group (s_{ij})

The car ownership data need in this study are the probabilities that households in various income groups own cars of different ages. These probabilities are denoted s_{ij}, where i refers to the income group and j to the car age group. Many s_{ij} estimates are required. With 487 geographic areas (243 SMSAs each with a central city and suburb and one non-SMSA group), seven income groups, six car age groups, and two analysis years, 40,908 separate estimates are needed.

Ideally these probabilities would be estimated using a behavioral model of car ownership decisions, data to estimate the various parameters of this model, and projections of the independent variables for 1980 and 1990. Developing an empirical model to predict detailed car ownership probabilities would be a major research undertaking. Also, all the necessary data does not now exist. The actual process used to estimate the s_{ij}s falls short of this ideal.

Estimates of the s_{ij}s in this study are calculated from two sets of 1970 data. Two data sets are required because no one set includes the necessary information. The first data are *conditional* probabilities referred to as s_{ijk}, which is the probability that a household in income group i owns a car in age group j, *given* that it own k cars. For k, three groups are used—one car, two cars, and three or more cars. These conditional probabilities are assumed to be common to all SMSAs. The second data are s_{ik}, estimates of the probabilities that households in various income groups own one, two, or three or more cars. These data were obtained separately for all the geographic areas in the model. The required car ownership information, the s_{ij}s, were calculated from these two data sets by multiplying each conditional car ownership probability by the corresponding probability of owning one, two, or three or more cars, and then summing the three figures. This calculation is shown in Equation 2.16.

$$s_{ij} = \sum_{k=1}^{3} (s_{ijk} \cdot s_{ik}) \tag{2.16}$$

where: s_{ij} = probability that a household in income
group i owns a car in car age group j

s_{ijk} = probability that a household in income
group i owns a car in car age group j
given that the household owns k cars

s_{ik} = probability that a household in income
group i owns k cars

Examples of the two kinds of car ownership data used in these calculations are given below.

The s_{ijk}s. Table 2-1 presents estimates of the s_{ijk}s for the six car age categories, seven income groups, and three car ownership levels used in this study. These estimates are derived from data provided by the Michigan Survey Research Center.[7] For example, the first entry, .014, indicates that there is a .014 probability that a household earning less than \$3,000 will own a new car, *given* that it owns one car. The age of car probabilities add up to 1.0 for one car households and up to 2.0 for two car households. These probabilities may add up to more than 3.0 for the last car ownership category since the average car ownership may be more than three cars per household.

The conditional car ownership probabilities in Table 2-1 may in fact differ across geographic areas. It is impossible to compute reliable estimates for urban areas or to test statistically for differences among particular urban areas because of the small number of observations (3,000 households in the total national sample). But these conditional probabilities were compared for different regions of the country and for different city and suburban population size categories. Statistically significant differences in these probabilities among regions and city size classes are very infrequent.[d] These results suggest that inter-

---

[d]The comparisons consisted of calculating chi-square significance levels for variations in conditional car ownership probabilities by the four regions and the six population size categories identified in the Michigan data. The significance levels can be interpreted as the probabilities that the differences in car age distributions for regions or for population size groups are due to chance. Of the 38 cells with enough data to perform the chi-square calculation, only three had significance levels below 0.05, the ordinary cutoff point for statistical significance in this kind of test. For almost all these estimates, therefore, there is a very high probability that the differences by region or by size group occurred by chance and that the underlying car age distribution (after accounting for income and number of cars owned) is the same.

urban variations are not large. The estimates in Table 2-1 therefore seem reliable estimates of the s_{ijk} probabilities for all urban areas.

The s_{ik}s. The 1970 Census of Population and Housing provides data on these car ownership probabilities.[8] The published census data provide

Table 2-1. Conditional Probabilities of Car Ownership by Car Age, Income Group, and Number of Cars Owned

Car Age	\<3	3-5	5-7	7-10	10-15	15-25	25+
				Income Group			
				One Car			
New	.014	.034	.050	.023	.040	.093	.0
1	.028	.069	.068	.144	.176	.157	.266
2	.070	.063	.086	.095	.116	.164	.133
3-4	.142	.190	.276	.248	.300	.343	.234
5-7	.276	.362	.319	.286	.294	.193	.300
8+	.470	.282	.201	.205	.074	.050	.067
Total	1.0	1.0	1.0	1.0	1.0	1.0	1.0
				Two Car			
New	.0	.100	.002	.034	.066	.090	.213
1	.154	.100	.032	.114	.257	.324	.320
2	.077	.150	.290	.200	.217	.212	.340
3-4	.461	.200	.324	.340	.460	.502	.596
5-7	.308	.550	.580	.730	.558	.542	.404
8+	1.000	.900	.742	.582	.442	.330	.127
Total	2.0	2.0	2.0	2.0	2.0	2.0	2.0
				Three Car			
New	.0	.5	.429	.141	.093	.048	.117
1	.0	.5	.143	.141	.281	.237	.529
2	.0	.0	.147	.141	.250	.378	.412
3-4	.0	1.0	.571	.489	.625	.684	.647
5-7	2.0	.0	.428	.699	.843	.993	.706
8+	1.0	1.0	1.428	1.40	1.031	.660	.706
Total	3.0	3.0	3.146	3.011	3.123	3.0	3.117

SOURCE: Author's analysis of the following data: University of Michigan Institute for Social Research 1970 Survey of Consumer Finances (Ann Arbor, Michigan: University of Michigan, nd).

information for all SMSAs and for the non-SMSA group on the number of cars owned, crosstabulated by income group, by total SMSA and central city residence, and by owner and renter status. Table 2-2 lists the probabilities of owning one car, two cars, or three or more cars for three SMSAs—Boston, Los Angeles, and Topeka. The last row gives the average cars per household for each group. This table shows the nature of this data and also indicates the wide variation in automobile ownership rates among different segments of the American population.

Vehicle Miles of Travel by Car Age and Income Group (V_{ij})

The auto travel data required are similar to the car ownership data. Estimates of annual auto miles driven must be given by income group and car age group for all the geographic areas in the study. The formula used to calculate average mileage driven is the following:

$$V_{ij} = \sum_{k=1}^{3} (V_{ijk} \cdot s_{ik}) \tag{2.17}$$

where: V_{ij} = annual vehicle miles of travel driven by a household in income group i in cars of car age group j

V_{ijk} = annual vehicle miles of travel for a household in income group i driven in a car in car age group j, *given* that the household owns k cars

s_{ik} = probability that a household in income group i owns k cars

The key to these estimates are the V_{ijk}s.[9] The car ownership data, the s_{ik}s, are the same used to calculate car ownership probabilities.

The V_{ijk}s are conditional estimates of miles driven, *given* that the household owns one, two, or three or more cars. These conditional data are listed in Table 2-3. As with the conditional car ownership data, it is assumed that these travel data apply to households in all geographic areas. Unlike car ownership probabilities, geographic differences in miles traveled cannot be examined, because there is no geographic detail in the available data. But geographic differences in miles of travel are probably small, after the household's income, the number of cars owned, and the car's age are accounted for.

As Equation 2.17 shows, the final travel estimates by car age and income group are obtained by multiplying the conditional estimate times the probability of owning one, two, or three or more cars and then summing over all car ownership groups. This procedure is analogous to that used to calculate the final car ownership probabilities.

Table 2-2.   Probabilities of Owning One, Two, and Three or More Cars in Boston, Los Angeles, and Topeka by Income Group and Central City or Suburban Residence

		<3	3-5	5-7	7-10	10-15	15-25	25+
				Income Group				
				Boston				
CC	1 car	.18	.28	.40	.57	.66	.59	.47
	2 car	.02	.03	.04	.06	.12	.25	.31
	3+ car	.01	.01	.01	.00	.01	.04	.10
	c/HH	.25	.35	.51	.70	.91	1.19	1.39
SUB	1 car	.35	.52	.62	.69	.60	.37	.31
	2 car	.06	.07	.10	.17	.32	.52	.58
	3+ car	.01	.01	.01	.01	.03	.10	.20
	c/HH	.49	.69	.87	1.08	1.33	1.69	1.96
				Los Angeles				
CC	1 car	.37	.53	.62	.60	.45	.25	.16
	2 car	.07	.10	.16	.26	.44	.57	.56
	3+ car	.01	.01	.02	.03	.07	.17	.26
	c/HH	.55	.78	1.00	1.23	1.54	1.88	2.04
SUB	1 car	.46	.61	.66	.58	.39	.21	.14
	2 car	.10	.14	.21	.33	.51	.58	.55
	3+ car	.02	.02	.02	.04	.09	.20	.29
	c/HH	.70	.95	1.15	1.36	1.66	1.98	2.13
				Topeka				
CC	1 car	.45	.60	.65	.56	.37	.23	.15
	2 car	.06	.15	.24	.34	.53	.57	.49
	3+ car	.01	.03	.03	.05	.10	.20	.29
	c/HH	.61	.97	1.20	1.38	1.71	1.95	2.02
SUB	1 car	.61	.74	.64	.51	.32	.23	.15
	2 car	.12	.18	.31	.45	.56	.62	.52
	3+ car	.03	.02	.01	.06	.09	.19	.22
	c/HH	.94	1.16	1.30	1.60	1.72	2.06	1.84

SOURCE: U.S. Bureau of the Census, *Census of Housing: 1970 Metropolitan Housing Characteristics,* Final Report HC(2)-30, HC(2)-120, HC(2)-222 (Washington, D.C.: U.S. Government Printing Office, 1972).

Table 2-3. Expected Vehicle Miles of Travel by Car Age, Income Group, and Number of Cars Owned

Car Age	<3	3-5	5-7	7-10	10-15	15-25	25+
				Income Group			
				One Car			
New	245	595	875	403	700	1,628	0
1	409	1,007	993	2,102	2,570	2,292	3,884
2	882	794	1,083	1,197	1,449	2,066	1,676
3-4	1,470	1,967	2,856	2,567	3,105	3,550	2,412
5-7	2,594	3,402	2,999	2,679	2,763	1,814	2,820
8+	3,650	2,191	1,568	1,599	562	390	523
Total	9,250	9,956	10,374	10,547	11,149	11,740	11,315
				Two Car			
New	0	1,770	566	602	1,168	1,593	3,770
1	2,680	1,740	557	1,983	4,472	5,638	5,568
2	1,040	2,025	3,915	2,700	2,930	2,862	4,590
3-4	5,578	2,420	3,920	4,114	5,566	6,074	7,212
5-7	2,957	5,280	5,568	7,008	5,357	5,203	3,878
8+	8,900	8,010	6,604	5,180	3,934	2,937	1,130
Total	21,155	21,245	21,130	21,587	23,427	24,307	26,148
				Three Car			
New	0	8,550	7,336	2,411	1,590	821	2,000
1	0	8,350	2,388	2,355	4,693	3,958	8,034
2	0	0	2,014	1,932	3,425	5,179	5,644
3-4	0	14,900	8,508	7,286	9,313	10,192	9,640
5-7	21,600	0	4,622	7,549	9,104	10,724	7,625
8+	8,500	8,500	12,138	11,900	8,763	5,610	6,001
Total	30,100	31,750	37,006	33,433	36,888	36,484	39,744

SOURCE: Author's analysis of the following two data sets: University of Michigan Institute for Social Research, *1970 Survey of Consumer Finances* (Ann Arbor, Michigan: University of Michigan, nd); and U.S. Department of Transportation, *Nationwide Personal Transportation Study*, Report Number 2, *Annual Miles of Automobile Travel* (Washington, D.C.: U.S. Department of Transportation, April 1972)

LIMITATIONS OF THE PROCEDURES

To analyze a complicated public policy requires a great deal of information on the nature of the policy as well as its surroundings. The information available is invariably incomplete. The precise character of the policy is usually not known.

The world into which a given policy is born is also too complex to capture all the ramifications of interest to policy makers. Designing a procedure to estimate the economic impacts of the current federal automobile emission control program and several plausible alternatives is, therefore, a difficult and inherently incomplete task.

The procedures in this study attempt to isolate the major determinants of the costs and benefits of auto emission control, to provide a logical framework for estimating the costs and benefits to households in different categories, and to use the available data to put empirical meat on these structural bones. Some discussion of the limitations of this procedure is beneficial, if only as a caveat on the conclusions which are drawn from it.

There are two major types of limitations: (1) specific inadequacies in the analysis or empirical estimation; and (2) general limitations on the scope of the analysis. The first sort of limitation relates to what *was done* in the process of generating empirical estimates and may be a defect in the statistical technique used to estimate parameters, a missing body of data, or an assumption made which is unrealistic but incorporated in the procedure because no better assumption can be used. The second type of limitation usually relates to what *was not done* or not attempted in the study. Usually these latter limitations derive from the basic approach of the study.

Some specific limitations relating to the car ownership and car use estimates were discussed in the last section. Among the other specific problem areas encountered in the estimation process are: determination of emission cost penalties, the new car price-cost relationship, and car depreciation rates (Chapter Three);the estimation of vehicle miles of travel by car age and income group (Chapter Four); and the estimation of the relationship between pollutant emissions and concentrations (Chapter Five). All of these difficulties are discussed in later chapters, and there is no reason to duplicate those discussions here.

Estimating the various parameters in the model involves the use of several data sets, including tens of thousands of numbers. Given this size, it is inevitable that some errors and omissions are present in the data. Great care was taken to test the results at various stages for accuracy and internal consistency, but certainly not all of the errors were detected. However, the basic conclusions do not appear to be very sensitive to minor errors which certainly remain in the data.

At several points in the empirical analysis decisions had to be made which would affect the results differently for households in different income groups. It is difficult to determine the net effect of these potential biases when some favor lower income groups and others favor higher income groups. To avoid this ambiguity, the study maintains a consistent bias in one direction. In selecting parameters and in making assumptions on the cost side, care was taken consistently to *understate* the cost burden to lower income groups. Similarly on the benefit side, care was taken to *overstate* the benefits to lower income groups.

Thus the costs may fall more heavily on the poor and the benefits may accrue proportionately less to the poor than the results in this study indicate.

The general limitations of an empirical study are usually less discussed than its specific inadequacies. This economic study focuses on the distributional impacts of auto emissions control and includes estimates relating to the economic efficiency impacts. But there are many noneconomic matters which are excluded from its purview. For example, no attention is paid to the administrative, legal, or political ramifications of federal automobile emission controls. In addition, decisions to alter the automobile emission control program may depend on comparisons with other policies to improve health and welfare. These alternatives are not evaluated in this study.

A complete multidisciplinary analysis incorporating all relevant alternatives would be extremely difficult and probably quite unwieldy. Indeed, it may well be that policy makers are better served by careful studies which limit their focus rather than by very general studies that attempt to cover all aspects of complicated matters. But this limited perspective should be kept in mind in interpreting the conclusions of this or any other study.

NOTES TO CHAPTER TWO

1. U.S. Environmental Protection Agency, *Annual Report of the Administrator of the Environmental Protection Agency* (Washington, D.C.: U.S. Government Printing Office, October 1973).
2. Chase Econometrics Associates, Inc., *Phase II of the Economic Impacts of Meeting Exhaust Emission Standards 1971–1980* (Springfield, Virginia: National Technical Information Service, December 1971); Ad Hoc Committee on the Cumulative Regulatory Effects on the Cost of Automotive Transportation (RECAT), *Final Report* (Washington, D.C.: U.S. Government Printing Office, February 1972); H.D. Jacoby and J.W. Steinbruner, *Clearing the Air* (Cambridge, Mass.: Ballinger Publishing Company, 1973). These national studies have other features, such as including a learning curve for the automobile industry so that estimated per car control costs decline over time or taking into account the uncertainty of actual emissions under various control systems.
3. Gregory K. Ingram and Gary R. Fauth, *TASSIM: A Transportation and Air Shed Simulation Model, vols. I and II,* Final Report to the U.S. Department of Transportation under Contract DOT-OS-30099 (Springfield, Va.: National Technical Information Service, May 1974).
4. See F.C. Wykoff, "A User Cost Approach to New Automobile Purchases," *Review of Economics Studies,* XL (July 1973): 377–390.
5. U.S. Environmental Protection Agency, *Supplement Number 2 for Compilation of Air Pollutant Emission Factors,* Report AP–42, 2nd ed. (Washington, D.C.: U.S. Government Printing Office, September 1973).

6. This technique has been developed by the federal EPA, although the basic concept is used widely. See U.S. Environmental Protection Agency, *Modified Rollback Computer Program Documentation* (Washington, D.C.: Environmental Protection Agency, November 1973).

7. Each year the Michigan Survey Research Center surveys approximately 3,000 families throughout the country on a wide range of subjects, one of which is their car ownership characteristics. For a description of this data, see George Katona et al., *1970 Survey of Consumer Finances* (Ann Arbor, Michigan: University of Michigan, 1971).

8. U.S. Bureau of the Census, *Census of Housing: 1970 Metropolitan Housing Characteristics*, Final Report HC(2)-1 to HC(2)-244 (Washington, D.C.: U.S. Government Printing Office, 1972).

9. This VMT data is obtained from the following study: U.S. Department of Transportation, *Annual Miles of Automobile Travel*, Report No. 2 of the Nationwide Personal Transportation Study (Washington, D.C.: U.S. Government Printing Office, April 1972), p. 8.

Chapter Three

Emission Control Technology and Costs

Much of the discussion of the federal automobile emission standards has centered on the technology needed to control emissions. The first question addressed was whether it was *possible* to control automotive emissions to the low levels mandated by the Congress in the 1970 Clean Air Act. This factual question in turn raised a host of subsidiary technical and legal questions. How are emissions to be measured? Do the standards apply to averages for the model year or does every car produced have to be below the standard? Should prototype cars or assembly line cars be tested to judge compliance with the standards? How should the durability of controls as the car is driven be judged?

Although many of these subsidiary questions are still not completely answered, by now there is a general consensus that the control levels mandated by the Congress are technologically possible, given reasonable answers to the various technical and legal questions. There is still some controversy about meeting and maintaining the most stringent control level for NO_x, but even there most commentators agree that the level can be achieved at some costs and at some time. Indeed, most of the disagreement now concerns which is the best technology to meet the standards, what is the time frame for developing alternative technologies, and what are the costs and other penalties associated with the various control technologies.

The purpose of this chapter is to derive estimates of the cost penalties under the current program in model year cars from 1971, when controls under the 1970 Clean Air Act began, to 1990, the last year in the analysis. Deriving these estimates requires characterizing the emission control technology employed on cars over this 20 year period and specifying the impact of these technologies on car production and development costs, fuel economy, and repair and maintenance expenses.

Although technology and cost information for future years is not certain, enough is known about emission control technology to make reliable estimates. Much of this technology and cost information has been collected and developed by the Committee on Motor Vehicle Emission (CMVE) of the National Academy of Sciences. The CMVE was established under a grant from the U.S. Senate to the Academy to study these and other related matters. The estimates in this study are derived primarily from the CMVE estimates reported in February 1973.[1] While some of these estimates may eventually be revised, they seem to represent reasonable approximations to the likely technology and costs of automobile emission control over this study's time horizon.

EMISSION CONTROL TECHNOLOGY AND PER CAR COSTS

The starting point for estimating emission control technology and costs for model years 1971 to 1990 is to consider the emission control standards in force for each year. Table 3-1 lists the standards (in grams per mile) that were promulgated by EPA under the 1970 amendments for each model year. All standards are expressed in terms of the latest federal procedure to measure auto emissions—the Constant Volume Sampling Procedure (CVS-CH). This measuring procedure uses a 23 minute driving cycle developed to simulate average driving in downtown Los Angeles.[2] Table 3-1 also lists the baseline emissions for 1970 model year cars.

The differing standards make it clear that vastly different control technology will be needed for the various model years in our study. From the

Table 3-1.   Federal New Car Exhaust Emission Standards

Model year	HC (g/mile)	CO (g/mile)	NO_x (g/mile)
1970	4.1	34.0	NR
1972	3.0	28.0	NR
1973	NC	NC	3.1
1975[a]	1.5	15.0	NC
	(0.9)	(9.0)	(2.0)
1976	0.41	3.4	2.0
1977	NC	NC	0.4

SOURCE: J.B. Heywood, "Impact of Emission Controls: 1968–1974," in F.P. Grad et al., *The Automobile and the Regulation of Its Impact on the Environment* (Norman, Oklahoma: University of Oklahoma Press, 1975).

[a]Standards in parentheses are the 1975 standards for California.

NR-no requirement.
NC-no change.

1970 baseline to the 1976 model year, allowable emissions of CO and HC are reduced by a factor of ten. Allowable emissions of NO_x are reduced by a factor of almost 13 from their uncontrolled level in 1970 to the final controls required in 1977 model year cars. Of course emission control technology may change even after the auto standards stabilize in model year 1977. Moreover, since this study is ultimately interested in the *costs* of meeting the standards in each model year, there is the possibility that the costs of using the same technology will decline over time.

As noted above, the technology assessments and cost estimates used in this study are based primarily on data from reports issued by the CMVE. While the original mandate of the CMVE was to investigate the technological feasibility of meeting the motor vehicle emissions standards prescribed in the 1970 Clean Air Act Amendments, in its reports the CMVE provided informed judgments on the likely emission control systems to be used to meet the various standards, the feasibility of mass producing promising systems, the projected performance of such emission control systems in customer usage, and the costs per vehicle of acquiring, maintaining, and operating the emission control systems.

Table 3-2 shows the per car production costs of emission control

Table 3-2. Per Vehicle Emission Control Costs for Model Years 1971 to 1990

Model Year	Per Vehicle Emission Control Costs ($)
1971	10
1972	10
1973	50
1974	50
1975	140
1976	140
1977	265
1978	265
1979	265
1980	265
1981	176
1982	176
1983	176
1984	176
1985	176
1986	176
1987	176
1988	176
1989	176
1990	176

Note: Costs are measured in 1972 dollars throughout this study

modifications by model year predicted in this study. The rest of this section discusses the technological assumptions and other judgments which lie behind these cost estimates. The pattern of emission control technological development under the 1970 Clean Air Act Amendments splits up into three time periods: (1) 1971 to 1976, (2) 1977 to 1980, and (3) 1980 to 1990.

1971 to 1976

The emission control systems developed for the six model years from 1971 to 1976 provide increasingly strict control of all three pollutants, culminating in the final emission control targets for both HC and CO. Because of the relatively long lead times in automobile production and, perhaps, because of technological conservatism on the part of the auto industry, the typical emission control modifications over this period consist of modifications to the standard internal combustion engine (ICE).

The required emission reduction in the model years from 1971 to 1974 was achieved for the most part by minor changes in engine design and more substantial changes in engine operating conditons.[3] Since NO_x standards did not begin until the 1973 model year, most changes over this period were designed to reduce HC and CO, which occur under roughly the same circumstances. Most automobile companies achieved compliance by having engines operate with leaner fuel-air mixtures and faster acting chokes. These modifications in engine operating conditions required some minor engine changes to counteract the engine durability problems resulting from higher temperatures in the engine and exhaust system. Control of NO_x emissions was actually made more difficult by some of these measures since the higher combustion temperatures which retard HC and CO increase NO_x emissions.

The major mechanism employed to control NO_x from 1973 to 1976 model year cars is exhaust gas recycling (EGR). With EGR some of the exhaust gas is recycled back into the combustion chamber thereby lowering the combustion temperature and inhibiting the formation of NO_x. There is some loss in drivability with the use of EGR. The car may stall more often, accelerate more slowly, and operate less smoothly. In addition, the fuel-air mixture must be enriched when EGR is used, and this enrichment imposes a fuel penalty. Some modifications in the 1975 and 1976 model cars compensate for the drivability disadvantages of EGR. For example, a new carburetor is designed to compensate for changes in air density with altitude which would otherwise result in drivability problems.

The major emission control task for 1975 and 1976 cars is to reduce HC and CO emissions to their final control levels of 0.41 g/mile and 3.4 g/mile respectively. Most manufacturers meet these standards by adding an oxidizing catalytic converter to the exhaust system of the standard ICE.[4] The basic function of the catalytic converter is simple. The exhaust gases pass through the mufflerlike converter where they undergo a chemical reaction which converts HC and CO to harmless gases and water. A platinum catalyst in the converter provides the impetus for these chemical reactions.

Considerable controversy surrounds the catalytic converter technology and its ability to obtain the low HC and CO levels required by the standards, particularly as the car is driven. The catalytic converter has certain inherent limitations, the major one being that emissions are not oxidized in the converter when the engine is just started and the platinum catalyst is cold (i.e., before it reaches its "light off" temperature). To reduce these "cold start" emissions, 1975 and 1976 cars include several modifications to preheat the air and fuel mixture and reduce the emissions generated during start-up.

Much of the concern about the catalyst system's effectiveness has been based on its alleged lack of durability.[5] It is well known that the catalyst is rendered ineffective if the fuel used contains catalyst poisons such as lead, phosphorus, and sulfur. The push to develop lead-free gasoline during this period is due primarily to this catalyst poison problem, although removal of lead from gasoline provides air quality benefits in its own right. Wide variations in the temperature and the fuel-air mixture in the exhaust gas may also impede the catalyst's effectiveness. In addition, the catalyst's potency deteriorates naturally with age.

It is hard for laymen to evaluate these arguments about the effectiveness and durability of emission control in catalyst equipped cars. The major American automobile companies, particularly General Motors, seem confident that these potential problems have been overcome and that the catalytic converter is effective and durable. The EPA estimates of new car emission rates and deterioration rates for 1975 and 1976 model year cars,[6] which are used to calculate automobile emissions in this study, imply that the catalyst system is effective in reducing new car emissions to the mandated levels and that emissions do not deteriorate markedly over time.

1977 to 1980

All cars manufactured in the 1977 model year and later must meet the final emission standards listed in Table 3–2, which involve reducing NO_x emissions to 0.4 g/mile. The four model years 1977 to 1980 are separated from the 1981 to 1990 decade because in the early years under the final standards the major American automobile companies will probably use different control technology. The technology most likely to be used by American manufacturers in the period from 1977 to 1980 is a modified version of the 1976 configuration called the dual catalyst system.[7] The dual catalyst system incorporates two catalytic converters, the oxidizing catalyst for HC and CO and a reducing catalyst for NO_x.

Estimating the costs, as well as the effectiveness, for the dual catalyst system is more difficult than for the previous emission control systems because the technology of the dual catalyst system is less developed and its production costs less documented. A supplementary CMVE report, however, estimated the costs of the dual catalyst system for seven body types ranging from subcompact to luxury.[8] These per car costs (over the 1970 base level car) range from $138 for a compact car to $491 for a luxury car. The average per car emission control cost for 1977 to 1980 model year cars, $265, was calculated using these CMVE estimates and the production mix by body type listed in Table 3–3.

All cost estimates for post-1977 cars are based on this mix of body types, which is concentrated in the small car end. It is projected that cars of intermediate size and below will account for 70 percent of car sales from 1977 model on. In 1971, these cars accounted for 45 percent of total sales. This prediction of an increased small car population is compatible with recent industry statements.[9] The increased demand for smaller cars is generally attributed to the recent rise in gasoline prices. It is possible that this shift to smaller cars is a temporary one, and that American car buyers will eventually return to large cars. In light of the long term prospects for higher fuel costs, however, the trend to smaller cars seems likely to be permanent.

1981 to 1990

While emission standards are the same for cars manufactured from 1981 to 1990 as for the previous four years, this study's estimate of average emission control costs declines by $90 from 1980 to 1981 and stays at the 1981 level for the rest of the decade. The drop in 1981 reflects the assumption that in 1981 two new control systems, the stratified charge system and the fuel injection system, will replace the dual catalyst carbureted system. Both of these technologies were identified by the CMVE as systems which may meet the 1977 emission control standards.[10] Both systems promise to reduce emissions to the

Table 3–3. Projected Fraction of Cars by Body Type for Model Years 1977 to 1990

Body Type	Projected Fraction of Total Sales	Actual 1971 Fraction of Total Sales[a]
Subcompact	.2	.07
Compact	.3	.08
Intermediate A	.1	.11
Intermediate B	.1	.19
Standard	.17	.28
Standard/Luxury	.1	.19
Luxury	.03	.08

[a]These data are from L.H. Lindgren, *Supplemental Report on Manufacturability and Costs of Proposed Low-Emission Automotive Engine Systems* (Washington, D.C.: National Academy of Sciences, January 1973), p. 3.

1977 levels at lower costs and with fewer drivability and fuel economy penalties than the dual catalyst system.

The fuel injection system includes an electronic fuel injection (EFI) mechanism to provide more precise fuel metering than the conventional carburetor. The EFI mechanism can reduce pollutant concentrations in the exhaust and still avoid some of the drivability and fuel economy penalties incurred by cars which use EGR and catalysts. The proper fuel mixture is determined by an oxygen sensor which provides feedback on the level of oxygen in the exhaust stream. To control emissions to the low levels required by the 1977 standards, this system will use a single catalyst to promote both oxidation of HC and CO and reduction of NO_x.

Several systems have been developed incorporating the stratified charge modification to the internal combustion engine. The most developed stratified charge emission control system, that of Honda of Japan, achieves charge stratification by adding a small precombustion chamber to the cylinder head and using two carburetors. A fuel-rich mixture is provided to the precombustion chamber by one carburetor to ensure good ignition. The main carburetor and fuel intake system feed a fuel-lean mixture to the main combustion chamber. This fuel-lean mixture produces fewer HC and CO emissions in the exhaust, and a slow burning flame in the combustion chamber reduces NO_x formation.

A supplementary CMVE report includes estimates of the costs of the fuel injection system and the stratified charge system for seven body types.[11] The cost estimate used in this study for the 1981–1990 period, $176 per car, assumes that the stratified charge system will be used on four and six cylinder cars and that the fuel injection system will be used in eight cylinder cars. The body type mix assumed for this calculation is the same as that used to calculate average control costs for model years 1977 to 1980.

This study assumes that these alternative technologies will be introduced in the 1980s rather than in the late 1970s because several developments must occur before each can be mass produced by American manufacturers.[12] For the electronic fuel injection system, a durable oxygen sensor for the feedback system must be developed, the EFI system must be improved to provide greater control of the air-fuel ratio during all engine operating conditions, and a sufficiently durable single catalyst must be found. Prototypes of the Honda stratified charge system have already met the 1977 standards, and Honda intends to market cars with stratified charge engines in 1977.[13] But several years are needed to transfer this technology to American manufacturers for mass production. These developments will probably not be accomplished until the 1980s.

This study may overstate emission control costs for several reasons. The changeover to less expensive emission control technology may be more rapid than predicted here. Some fuel injection and stratified charge emission control systems will be used on part of the car fleet before 1981. But given the long lead times necessary to introduce new technology into the production process and the likely reluctance of the major American manufacturers to write off their capital investment in the standard ICE technology, it is doubtful that a large share of cars will be equipped with these other technologies before 1981.

Emission control costs may also be lower in later years because of the learning curve effect on production costs for emission control technology.[14] Improvements over time in manufacturing procedures which lower average production costs are common for manufactured goods. This learning curve effect on average costs is particularly important in new products, which may have high production costs initially but decline in cost as companies lean cost-saving methods of production. If automobile companies are able to produce or install emission control equipment more cheaply in later years, the average cost of emission controls will be lower than predicted here.

Another factor which may lead to lower emission control costs in the 1980s is the use of control technology other than the fuel injection or stratified charge engines. Many alternatives have been proposed, ranging from exotic steam cars, electric cars, and cars that run on alcohol, to more likely cars equipped with diesel engines and Wankel rotary engines. None of these alternatives is now capable of meeting the 1977 emission control standards, although some may in the 1980s.[15] But even if the technology mix in the 1980s is different than the one predicted here, the results of this study are not likely to be substantially affected. Engine technology is only important to this study insofar as it influences the costs of emission control in each model year. Different emission control technology from that predicted here will probably not reduce the per car cost of meeting the final 1977 standards much below the $176 per car used in this study.

NEW CAR PRICE INCREASES

Costs incurred by automobile companies to develop and install emission control technology on their cars will be passed on to households when these costs are reflected in higher new car prices. The problem is to predict the fraction of added per car costs passed on to new car buyers. This problem can be stated mathematically as follows:

$$P_t = a C_t \tag{3.1}$$

where: P_t = average increase in selling price due to emission control costs for new cars manufactured in model year t

a = fraction of cost increase passed on as price increase

$$C_t = \text{average increase in per vehicle cost due to emission controls for new cars manufactured in model year } t$$

In this section an estimate of a is chosen and new car price increases are then estimated using the cost increases estimated in the previous section.[a]

The value of a used in this study to estimate selling price increases was obtained from a recent study of the automobile industry prepared by Chase Econometric Associates, Inc.[16] Chase assumed that the automobile industry behaves with respect to pricing as if it were a single monopolist. Separate demand equations were estimated for five categories of cars ranging from subcompact to luxury. The problem posed for the monopolist in this situation is to choose five new car prices simultaneously so as to maximize total industry profits. The emission control cost increase was viewed as a tax levied on each car category, and statistical analysis of time series data was used to estimate the response of the car prices to this tax. The estimates of the price-cost ratio obtained by Chase for the five car categories are close to one another and average 0.90. This average figure of 0.90 is used here for an estimate of a. The price increases predicted for each model year using Equation 3.1 and $a = 0.90$ are listed in Table 3–4.

These price increases imply a decline in automobile company profits (relative to the uncontrolled situation) because the price increase is only nine-tenths of the per car cost increase. Roughly one-half of the pretax profit drop will be borne by federal taxpayers because the auto companies will pay less corporate income tax. However, the magnitude of the fall in profits is small. In 1980, the per car profit decline is estimated to be $26.50 per car, which, assuming that ten million cars are sold, results in a $265 million drop in automobile company profits. For 1990, the profit decrease calculated in the same manner is $176 million. These profit changes are a small part of the total auto emission control costs, including the fuel economy losses and the maintenance cost increases estimated later in this chapter. Even for the automobile companies, they are modest. Using the industry rule of thumb of an average profit of 10 percent of list price and an average list price of $3,300, the decrease in profits in 1980 will be less than 8 percent. While these figures are crude, they indicate the modest nature of these profit decline estimates.

---

[a]One estimate of a was implicit in the CMVE's estimates of list price changes due to emission control costs. The CMVE's list price increases were obtained by adding a dealer markup and a manufacturer's profit to the per unit cost of emission controls. Using a dealer's markup of 22 percent of list price and a manufacturer's profit of 10 percent of list price, the CMVE's cost plus technique yields an estimate for a, the ratio of price to cost, of 1.47. The National Academy of Sciences, *Report*, pp. 90–93. The CMVE figure was not used in this study because the industry rules of thumb on which it is based relate to *list price* (often referred to as sticker price) changes, not the *selling price* changes with which this study is concerned.

The actual adjustments in the automobile market and related markets are probably more complicated than assumed here. The automobile companies have several options other than simply changing their prices for new cars. One possibility is to increase replacement parts prices. The demand for replacement parts is relatively insensitive to price changes, and the automobile companies could exploit this condition to maintain traditional profit rates.

Other economywide adjustments may result from emission control costs. For example, the demand for various factors of production may increase or decrease as a result of the input requirements of the control technology used. To take the clearest case, the demand for platinum increases when the oxidizing catalytic converter is used to control HC and CO emissions. This demand may increase the price of platinum, thereby generating windfall benefits to owners of that material and higher costs for emission control.

Except for the impact on the used car market considered in the next section, these other adjustments following automobile production cost increases are not considered in this study because they are likely to be small in magnitude. The distributional impacts of these complicated adjustments are minor compared to the impacts of increased automobile ownership and operating costs. The general conclusions of the study thus should not be affected by neglecting these complications.

Table 3–4.   New Car Price Increases Due to Emission Control Costs for Model Years 1971 to 1990

Model Year	Price Increase ($)
1971	9
1972	9
1973	45
1974	45
1975	126
1976	126
1977	238
1978	238
1979	238
1980	238
1981	158
1982	158
1983	158
1984	158
1985	158
1986	158
1987	158
1988	158
1989	158
1990	158

USED CAR PRICE INCREASES

Increases in new car prices indicate how control costs are divided among households, automobile company stockholders, and federal taxpayers. To estimate the costs borne by households in different income groups, it is also necessary to know the increase in used car prices resulting from these new car price increases. The price increase borne originally by the new car price purchaser will be shifted partly to used car purchasers. Since the fraction of the new price increase passed on to used car buyers varies with the age of the car, it is necessary to estimate the complete pattern of price increases by car age. These price increases by car age can then be combined with estimates of the car ownership by car age profiles of households to predict the distributional effects of the new car price increases.

The relationship between new car prices and prices of used cars of various vintages is complicated. There are two major effects, the depreciation effect and the substitute good effect. The depreciation effect is analogous to the standard notion of depreciation for the car as a whole. Like any durable good, the automobile deteriorates over time as parts wear out and maintenance costs increase. Because part of the value of a car stems from its style, obsolescence includes changes in the aesthetic appeal of cars. This physical and aesthetic deterioration is reflected in the declining price of the car as it ages. A similar decline for each model year should apply to the new car price increase due to emission control costs.

The substitute good effect is the used car price increase due to *current year* new car price rises. New cars and used cars of various ages are partial substitutes for one another. A potential car owner can choose a wide range of car age groups, although older cars are eventually scrapped when increased maintenance and repair costs far outweigh the price difference between the old car and other cars. If the price of new cars goes up, more car buyers will want to buy used cars, and this increased demand for a more or less constant stock will result in increased used car prices. Rises in used car prices obtained in this fashion are what are referred to as substitute good effects.

These two influences are illustrated graphically in Figure 3-1, where the price of a 1974 car is charted over time. These curves are designed only to illustrate the two effects on used car prices and have no empirical basis.

The lower curve in Figure 3-1 illustrates the price history of the 1974 car if there were no emission control costs. According to the illustrative numbers given, the average price of a 1974 car would decline from $3,000 in 1974 to $2,000 in 1977. But with emission controls added, the 1974 car is assumed to have a price of $3,300 when it is new. The solid line above the lowest curve illustrates the price change of this emission controlled 1974 car over time. The increase in price at each age is the depreciation effect of emission controls on used cars. For example, in 1977 the controlled price of a three year old car is $2,000 + *ab* in this illustration. Thus a household buying a three year old car in 1977 would pay an additional *ab* for emission control costs.

The dashed line in Figure 3-1 includes the substitute good effect as well as the depreciation effect. For example, suppose that the new car price of a 1977 car rose by an additional $200 because of added emission control costs. In 1977, when consumers are deciding among cars of various vintages, the $200 price rise for new cars will further elevate the price of the 1974 model year car. In this example, the price is *bc* greater as a result of this substitute good effect. A purchaser of the 1974 car in 1977 will therefore pay an additional *bc* because of emission controls on 1977 model year cars, above the *ab* paid because of controls on the 1974 car itself.

This substitute good effect generates windfall profits to sellers of used cars. For example, if a household owned a 1974 car in 1977, it could sell the car for an additional *bc* dollars over its expected sale price (assuming that it bought the car when new). This figure would represent a one shot gain, since the next owner would obtain no similar advantage. Note that this substitute good effect only operates when the price of new cars changes considerably. If new car prices are stable, the relative prices of new and used cars are determined only by the depreciation or aging effect.

Substitute good effects are ignored in this study because they will be quite small in 1980 and nonexistent in 1990. In 1990, this effect will not be present because it is estimated that new car prices in 1990 will be the same as those in the past nine years. Any windfall profits for sellers of used cars will have been exhausted by 1990. Some windfall gains may be obtained in 1980 for

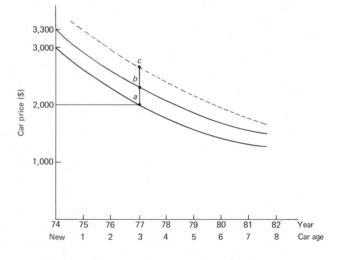

Figure 3-1. Influence of Emission Control Costs on the Automobile Depreciation Pattern

sellers of older cars, but they will be minor since new car price increases are assumed to be the same for model years 1977 to 1980.

Windfall gains will occur in the period from 1971 to 1977, when new car prices change substantially from year to year. But including these effects would not contribute significantly to this analysis. Spread over many years, these windfall gains would effect relatively small changes in the distribution of costs. However, if lower income groups are net sellers of used cars in this period, this is one factor working to lessen the relative burden of emission control costs on low income households. This possible bias is contrary to the general rule in this study of biasing results toward understating the relative burden to low income groups.

The used car price increases estimated in this study, therefore, depend only on the depreciation effect. The empirical estimates of these depreciation increases are derived from a recent study of automobile depreciation patterns by Griliches and Ohta.[17] Assuming the standard negative exponential form for the depreciation pattern, Griliches and Ohta estimated the depreciation rate to be 0.25. The formula used in this study to estimate the change in used car prices resulting from new car price increases is the following:

$$U_t = U_o \, e^{-.25t} \qquad\qquad (3.2)$$

where: U_t = increase in used car price

         U_o = new car price increase

         t = car age

MAINTENANCE AND FUEL COST INCREASES

The technology used by automobile manufacturers to control emissions results in some deterioration in fuel economy and some increase in repair and maintenance expenses. These penalties vary a great deal depending upon the emission control technology employed and therefore upon the model year. Table 3-5 lists the annual maintenance penalties and per mile fuel cost penalties estimated in this study for 1971 to 1990 model year cars. As with the hardware cost penalties, the 1970 model year is taken as the baseline case. Thus the figures in Table 3-5 represent cost increases over the 1970 model year car.

Meeting the final standards in model years 1977 to 1980 results in the greatest cost penalties. For these years added maintenance costs are estimated at $25 per year and added fuel costs are .588¢/mile. Both increases are roughly five times as great as those in model years 1971 and 1972, the years with the smallest penalties. The large 1977-80 costs reflect the disadvantage of the dual catalyst system. In the 1980s, when stratified charge systems and fuel injection systems replace the dual catalyst system, the maintenance and operating

Table 3-5.   Operating Cost and Maintenance Cost Penalties for Model Years 1971 to 1990

Model Year	Operating Cost (¢/mile)	Maintenance Cost ($/year)
1971	.119	5
1972	.119	5
1973	.289	5
1974	.289	5
1975	.251	15
1976	.251	15
1977	.588	25
1978	.588	25
1979	.588	25
1980	.588	25
1981	.175	10
1982	.175	10
1983	.175	10
1984	.175	10
1985	.175	10
1986	.175	10
1987	.175	10
1988	.175	10
1989	.175	10
1990	.175	10

cost penalties are substantially reduced. The maintenance cost penalty is only $10 per year and the fuel cost penalty is only .175¢/mile in the 1980s. The remainder of this section discusses the derivation of the penalties listed in Table 3-5.

Fuel Cost Increases

Fuel economy penalties due to emission control are usually reported as percentage penalties relative to the uncontrolled (1970) fuel economy. The per mile cost increases presented in Table 3-5 were calculated from percentage penalties using the following formula:

$$O = \frac{d \cdot g}{F(1-d)} \tag{3.3}$$

where: O = fuel cost penalty (¢/mile)

g = gas price (¢/gallon)

F = fuel economy without emission controls (miles/gallon)

d = fuel economy penalty with emission controls (fraction of uncontrolled fuel economy)

These calculations assume that the price of gasoline is $.50 per gallon (in 1972 dollars) and that baseline fuel economy is 13.0 miles per gallon for model years 1971 to 1974 and 15.0 miles per gallon for model years 1975 and beyond. The greater baseline fuel economy for cars in the later years is due to the larger fraction of small cars which this study assumes for cars produced in those years.

Estimating the fuel economy penalty for each model year, the d in Equation 3.3, is a complicated chore. The miles per gallon a car gets depends upon many design and operating characteristics, including engine size, car weight, carburetion, optional accessories (particularly air conditioning), cold or hot engine operation, tire friction, aerodynamic drag, and average vehicle speed.[18] Separating out the independent penalty of emission control modifications from these other factors is difficult, particularly since the penalty may vary according to the engine design and the operating conditions.

The fuel economy penalties used in this study are derived from several reports which summarize the available evidence. The EPA has collected and reported data on average fuel consumption for model years 1970 to 1973.[19] This data indicates that smaller vehicles had better fuel economy in 1973 than they did in the 1970 baseline year, while larger vehicles showed substantial declines over the four year period. The average fuel economy (using 1972 sales fractions) declined from 12.4 miles per gallon in 1970 to 12.1 in 1971, 12.0 in 1972, and 11.7 in 1973. If it is assumed that these changes are due to emission control modifications, these figures indicate fuel economy penalties of 2.0 percent in 1971, 2.9 percent in 1972, and 5.8 percent in 1973.

A study by John Heywood of the Massachusetts Institute of Technology suggests that the true fuel penalties for these early years of emission controls may be higher than these average figures because fuel economy improvements for smaller cars in this period are misleading.[20] Many of the subcompact and compact cars had been recently introduced in 1970, and some fuel economy modifications would probably have been made in the absence of emission control requirements. Much of the improvement in fuel economy for these cars from 1970 to 1973 was due to improved fuel metering which had already been introduced in larger cars.

Heywood also reviewed the evidence on fuel economy for the catalyst emission control systems. For the 1975 system with engine modifications and a catalytic converter, Heywood reported estimates of fuel economy penalties (relative to the 1973 case) ranging from no change to a 4 percent penalty.[21] The 1976 requirement for lower NO_x does not seem to affect fuel economy.

The fuel penalties for the dual catalyst systems are more difficult to predict because there is less information available, even from prototype versions. Heywood reports that automobile manufacturers estimate the fuel penalty for the dual catalyst system to be about 10 percent compared to equivalent weight 1973 model vehicles.[22] This would correspond to roughly a 17 percent penalty relative to the 1970 baseline case. Data obtained by the CMVE from the automobile companies in 1972 indicated larger penalties than those reported by Heywood. For the dual catalyst system, the fuel penalties ranged from 22 percent for subcompacts to 33 percent for luxury cars. The penalties reported by the CMVE were considerably smaller for the fuel injection and stratified charge control systems; penalties for both systems were about 9 percent for all car sizes.[23]

Estimates of fuel economy penalties in this study are drawn from this limited information, taking into account that penalties should decline over time as technology is refined. It is estimated that the fuel economy penalty due to emission controls is 3 percent for 1971 and 1972 model years and 7 percent for 1973 and 1974 cars. These figures are slightly larger than the EPA estimates to take into account the possible bias in fuel economy penalties for smaller cars in those model years. For 1975 and 1976 cars it is assumed that there is no additional penalty over the 1973 and 1974 model year cars. For model years 1977 to 1980, when the dual catalyst system will be used to control emissions, a 15 percent decrease in fuel economy is predicted. This is approximately twice the penalty for the previous year, although it is considerably less than the estimates made by the industry in 1972. For the 1980s, when EFI and stratified charge technology will be used to control emissions, the fuel penalty is projected to be considerably smaller, only 5 percent compared to the 1970 baseline.

Maintenance Cost Increases

Virtually no data exists on the maintenance costs car owners actually incur because of emission controls. Although the CMVE discussed the general problem of maintaining emission control effectiveness as the car is driven, their reports presented no estimates of the expected maintenance expenses for pre-1977 control systems.[24] Some studies have reported cost estimates for state inspection-maintenance (I-M) programs, which monitor car emissions and require high emitters to be repaired.[25] The costs of state I-M programs are not included in this study's estimates of control cost burdens because the current federal auto emission control strategy does not require I-M. Chapter Eight, which evaluates several alternatives to the current scheme, estimates the costs of adopting stringent I-M programs for the whole country and for a small group of SMSAs.

The maintenance cost increase estimates in Table 3–5 are best viewed as relatively optimistic guesses at the actual maintenance expenses associated with emission controls for each model year. An additional maintenance cost of $5 per year is projected for model years 1971 to 1974. The hardware in these precatalyst systems is simple and the maintenance expenses should be modest. For the single HC-CO catalyst equipped cars manufactured in 1975 and 1976, additional maintenance expenses of $15 per year are predicted. Most of the extra cost is for the replacement of the catalyst. The maintenance expenses

Table 3-6. Summary of Emission Control Costs for Model Years
1971 to 1990

Model Year	Per Car Production Cost Increase ($)	New Car Price Increase ($)	Per Mile Operating Cost Increase (¢)	Annual Maintenance Cost Increase ($)
1970	baseline case			
1971	10	9	.119	5
1972	10	9	.119	5
1973	50	45	.289	5
1974	50	45	.289	5
1975	140	126	.251	15
1976	140	126	.251	15
1977	265	238	.588	25
1978	265	238	.588	25
1979	265	238	.588	25
1980	265	238	.588	25
1981	176	158	.175	10
1982	176	158	.175	10
1983	176	158	.175	10
1984	176	158	.175	10
1985	176	158	.175	10
1986	176	158	.175	10
1987	176	158	.175	10
1988	176	158	.175	10
1989	176	158	.175	10
1990	176	158	.175	10

Note: All costs and prices are in 1972 dollars.

for the dual catalyst system are expected to be $25 per year. This latter estimate is considerably lower than the CMVE estimates, which range from $36 per year for subcompact cars to $52 per year for luxury cars.[26] The CMVE projected that the catalyst would have to be replaced twice in five years, while the figure in this study assumes only one replacement. In addition, car owners are unlikely to perform some of the maintenance embodied in the CMVE estimates, since, if anything, these maintenance expenses would degrade vehicle drivability. The automobile companies will also probably improve the reliability of emission control hardware over time. For the stratified charge and fuel injection systems, maintenance costs are assumed to be only $10 per year.

In calculating maintenance penalties for car age groups in 1980 and 1990, this study assumes that the penalties in Table 3-5 apply for all years of a particular car's life. In fact, annual repair and maintenance costs are likely to increase with age. A single figure is used for all years because no information exists on changes in these emission control penalties with age.

SUMMARY OF EMISSION CONTROL COSTS

The federal automobile emission standards will increase several types of auto-mobile costs. Research and development costs, retooling expenses, labor costs, and material costs connected with emission control technology will raise average vehicle production costs. These added costs will result in higher new car prices which in turn lead to greater used car prices. Per mile fuel costs will be greater because emission controls degrade fuel economy. Average maintenance costs will be greater because control hardware may require repair or replacement and more frequent adjustment.

The cost estimates derived in this chapter for model years 1971 to 1990 are summarized in Table 3-6. While the actual costs incurred by car owners may differ somewhat from these figures, the estimates in Table 3-6 are based on the currently available evidence. The estimation procedures and additional data needed to translate these cost penalties into emission control cost burdens for particular household groups are given in the next chapter.

NOTES TO CHAPTER THREE

1. National Academy of Sciences, *A Report by the Committee on Motor Vehicle Emissions* (Washington, D.C.: National Academy of Sciences, February 12, 1973).
2. A description of the CVS-CH federal test procedure and a discussion of the various alternative test procedures are found in the following: J.B. Heywood, "Impact of Emission Controls: 1968–1974," in F.P. Grad et al., *The Automobile and the Regulation of Its Impact on the Environment* (Norman, Oklahoma: University of Oklahoma Press, 1975), pp. 115–123.
3. See National Academy of Sciences, *Report,* pp. 87–89; and Heywood, "Impact of Emission Controls," pp. 123–124.
4. J.B. Heywood, "Future Emission Control Technology," in F.P. Grad et al., *The Automobile and the Regulation of Its Impact on the Environment* (Norman, Oklahoma: University of Oklahoma Press, 1975), p. 291.
5. The problem of catalyst deterioration figures prominently in the following study: H. C. Jacoby and J. W. Steinbruner, *Clearing the Air* (Cambridge, Massachusetts: Ballinger Publishing Company, 1973). But see Heywood, "Future Emission Control Technology," pp. 283–291, for a different view.
6. U.S. Environmental Protection Agency, *Supplement Number 2 for Compilation of Air Pollutant Emission Factors,* Report AP-42, 2nd ed. (Washington, D.C.: U.S. Government Printing Office, September 1973), 3.1.2–6.
7. Heywood, "Future Emission Control Technology," p. 293.
8. LeRoy H. Lindgren, *Supplemental Report on Manufacturability and Costs of Proposed Low-Emission Automotive Engine Systems,* prepared for the Committee on Motor Vehicle Emissions (Washington, D.C.: National Academy of Sciences, January 1973), pp. 16–18.

9. Agis Salpukas, "Autos' Future: Cutback in Size and Frills," *New York Times,* March 30, 1975, p. 1.

10. National Academy of Sciences, *Report*, p. 4.

11. Lindgren, *Supplemental Report,* p. 16–18.

12. National Academy of Sciences, *Report,* pp. 97–98; and Heywood, "Future Emission Control Technology," pp. 307–309.

13. National Academy of Sciences, *Report,* pp. 59–60.

14. See Ad Hoc Committee on the Cumulative Regulatory Effects on the Cost of Automotive Transportation (RECAT), *Final Report* (Washington, D.C.: U.S. Government Printing Office, February 1972).

15. National Academy of Sciences, *Report,* pp. 104–112.

16. Chase Econometrics, Inc., *Phase II of the Economic Impacts of Meeting Exhaust Emission Standards 1971-1980* (Springfield, Virginia: National Technical Information Service, December 1971), p. III–14.

17. Makota Ohta and Zvi Griliches, "Automobile Prices Revisited: Extensions of the Hedonic Hypothesis" (Harvard Institute of Economic Research Discussion Paper Number 325, Harvard University, Cambridge, Massachusetts, October 1973).

18. T.C. Austin and K.M. Hellman, "Passenger Car Fuel Economy–Trends and Influencing Factors" (Paper number 730790, presented to Society of Automotive Engineers, September 1973).

19. Ibid., pp. 25, 30.

20. Heywood, "Impact of Emission Controls," pp. 133–135.

21. Heywood, "Future Emission Control Technology," p. 291.

22. Ibid., p. 294.

23. Lindgren, *Supplemental Report,* pp. 16–18.

24. National Academy of Sciences, *Report,* pp. 69–86.

25. J.B. Heywood, "Inspection/Maintenance and Retrofit of In-Use Automobiles," in F.P. Grad et al., *The Automobile and the Regulation of Its Impact on the Environment* (Norman, Oklahoma: University of Oklahoma Press, 1975), pp. 148–250.

26. Lindgren, *Supplemental Report,* pp. 16–18.

Chapter Four

Estimation of Emission Control Cost Burdens

The overview in Chapter Two identified four major ways in which households will bear the costs of federal automobile emission control: (1) as automobile owners, who must pay higher prices for new and used cars; (2) as automobile operators, who must pay more for gasoline, maintenance, and repairs; (3) as automobile company stockholders, who will receive lower corporate dividends; and (4) as taxpayers, who must make up for lower corporate tax revenues.

The purpose of this chapter is to explain in detail the data and methodology used to estimate these components for the current control scheme. For each component, equations summarizing the steps in the estimation procedure are first presented, all of which culminate in formulas for calculating the costs borne by different income groups. As mentioned in the overview, these formulas are the same for all geographic areas. The underlying cost data needed for these estimation procedures are then presented. Finally national estimates of average cost burdens for each income group are presented.

No geographic detail is provided in this chapter. The cost estimates presented for the various components are designed to show how the procedures and the data are used to generate specific cost estimates and also to provide a preview of the detailed presentation of results in Part II.

AUTOMOBILE OWNERSHIP COSTS

Chapter Two overviewed the procedures used to estimate the increased automobile ownership costs households in various income groups will bear. This process is summarized by the following equations:

probability of car ownership by car age

$$s_{ij}^t = \sum_{k=1}^{3} (s_{ijk}^t \cdot s_{ik}^t)$$ (4.1)

car ownership cost by car age

$$O_j^t = (P_j^t - P_j^{t+1}) + (r \cdot P_j^t)$$ (4.2)

ownership cost by income group

$$O_i^t = \sum_{j=1}^{6} (s_{ij}^t \cdot O_j^t)$$ (4.3)

where O = ownership cost burden

s = probability of car ownership

P = price increase due to emission controls

r = interest rate

i = income group

j = car age group

k = number of cars owned group

t = analysis year

Equations 4.1 and 4.2 generate the two sets of data needed in Equation 4.3 to calculate annual ownership cost burdens by income group. The probabilities that households in each income group own cars in various age groups, the s_{ij}^t s, are dervied from data on two other sets of car ownership probabilities, as Equation 4.1 indicates. These car ownership data were explained in Chapter Two. Equation 4.2 shows that the estimates of increased ownership costs for cars in various age groups, the O_j^ts, are derived as the sum of increased depreciation costs and increased finance costs. The data needed to calculate these costs for car age groups in 1980 and 1990 were developed in Chapter Three.

Table 4-1 lists the O_j^ts used in this study. Increased annual ownership costs are listed for cars in six car age groups in 1980 and 1990. This table shows that ownership cost penalties vary considerably for cars in different age categories, with added costs due to emission controls much greater for owners of newer cars than for owners of older cars. For 1990, the cost increase varies from $5.30 for an owner of a car eight years or older to $50.41 for an owner of a new car. The spread is even greater in 1980, when the annual increase in

Table 4-1. Increased Annual Depreciation Costs, Finance Costs, and Total Ownership Costs by Car Age Category in 1980 and 1990

Car Age	Depreciation Costs ($)	Finance Costs ($)	Total Owner Costs ($)
	1980		
New	52.21	23.80	76.01
1	41.02	18.58	59.60
2	32.20	14.48	46.68
3–4	17.51	7.95	25.46
5–7	4.01	1.80	5.81
8+	.12	.10	.22
	1990		
New	34.61	15.80	50.41
1	27.25	12.34	39.59
2	21.38	9.61	30.99
3–4	14.65	6.66	21.31
5–7	8.03	3.61	11.64
8+	3.64	1.66	5.30

ownership costs is $76.01 for an owner of a new car and only $0.22 for an owner of a car eight years or older.

These differences in ownership costs by car age reflect the exponential decline in the car price increase due to emission controls as the car ages. In addition, for 1980 the ownership cost differences by car age reflect the different initial new car price changes for different model years. For example, the initial price penalty estimated in Chapter Three for a 1980 model year car is almost four times the new car price increase for a 1975 model year car. When the initial price increase for a 1975 car is depreciated exponentially for five years, the difference in cost borne by owners of 1980 and 1975 cars in 1980 is quite large. For 1990, the initial new car price increase is the same for virtually all of the cars on the road.

Table 4-2 lists the national estimates of average annual ownership costs by income group in 1980 and 1990. The average dollar burden increases with household income, ranging in 1990 from $7.59 for households with income less than $3,000 to $46.68 for households with over $25,000. This trend reflects the fact that as income increases, households own more cars and newer cars. As discussed in Chapter Two, comparing the dollar cost for households in different income groups does not reflect differences in relative burden endured by households. A more familiar procedure is to compare the percentage of total income represented by the dollar burden for each group. This percentage is called the *relative cost burden.* Relative cost burdens are listed for

Table 4-2. National Annual Ownership Costs by Income Group in 1980 and 1990

Income Group	1980 Costs ($)	1980 Relative Costs (%)	1990 Costs ($)	1990 Relative Costs (%)
<3	6.76	0.38	7.59	0.42
3-5	14.75	0.37	14.13	0.35
5-7	20.01	0.33	18.39	0.31
7-10	24.01	0.28	21.98	0.26
10-15	36.30	0.29	30.62	0.24
15-25	46.02	0.23	37.85	0.19
25+	60.11	0.20	46.68	0.16
Inequality Ratio		1.9		2.6

each income group to the right of the dollar cost burdens in Table 4-2.[a]

The cost bearing pattern in Table 4-2 indicates that auto ownership costs will be more burdensome to lower income groups. In economists' terminology, the cost pattern is *regressive*. While the dollar burden increases with household income, these costs account for a steadily decreasing fraction of household income. Although lower income households on average own fewer and older cars than higher income households, they spend a higher proportion of their income on car ownership.

Table 4-2 also includes the 1980 and 1990 national inequality ratios, defined as the ratio of the relative cost burden for households in the lowest income group to the relative cost burden for households in the highest income group. Although the costs are generally lower in 1990, they fall more heavily on the poor in the later year. The dollar costs decrease from 1980 to 1990 for all income groups except the lowest group, but the inequality ratio is 1.9 in 1980 and 2.6 in 1990. A more regressive pattern occurs in 1990 because the relative ownership cost burden for owners of older cars as compared to newer cars is greater in 1990.

AUTOMOBILE OPERATING COSTS

The steps used to estimate increased auto operating costs for households in various income groups are summarized by the following equations:

---

[a]In computing relative burden throughout this study, the average income for the below $3,000 income group is set at $1,800 and the average income of the above $25,000 income group is set at $30,000. These figures correspond roughly to median incomes for these tail end groups derived from more detailed income grouping. [1]

probability of car ownership by car age

$$s_{ij}^t = \sum_{k=1}^{3} (s_{ijk}^t \cdot s_{ik}^t)$$ (4.4)

vehicle miles of travel by car age

$$V_{ij}^t = \sum_{k=1}^{3} (V_{ijk}^t \cdot s_{ik}^t)$$ (4.5)

operating costs by income group

$$U_i^t = \sum_{j=1}^{6} (a_j^t \cdot V_{ij}^t) + \sum_{j=1}^{6} (s_{ij}^t \cdot M_j^t)$$ (4.6)

where: U = operating cost burden

   s = probability of car ownership

   M = maintenance cost increase

   V = vehicles miles of travel

   a = per mile fuel cost increase

   i = income group

   j = car age group

   k = car ownership group

   t = analysis year

Equation 4.6 shows that operating cost penalties are the sum of increases in annual fuel costs and increases in annual maintenance costs. The annual maintenance costs borne by households in each income group are calculated from car ownership probabilities and maintenance cost increases using the same procedure as used to calculate ownership costs. Estimating increased annual fuel costs is more complicated because estimates of the average number of miles traveled by car age for each income group are required. The steps to obtain the miles traveled estimates in this study were explained in Chapter Two; Equation 4-5 summarizes these calculations.

The estimates of annual maintenance costs, the M_j^ts, and per mile fuel costs, the a_j^ts, are listed in Table 4-3. These figures were obtained from the estimates derived in Chapter Three. Maintenance and operating costs penalties vary a great deal for cars of different vintages in 1980. In 1980, the penalties for new cars are about five times the penalties for cars of 1972 vintage and older.

Table 4-3. Increased Annual Maintenance Costs and Per Mile Fuel Costs by Car Age Category in 1980 and 1990

	1980		1990	
Car Age	Maintenance ($)	Fuel (¢)	Maintenance ($)	Fuel (¢)
New	25	.590	10	.175
1	25	.590	10	.175
2	25	.590	10	.175
3-4	15	.251	10	.175
5-7	5	.289	10	.175
8+	5	.119	10	.175

The maintenance and fuel costs penalties are the same for all car age categories in 1990 because it is assumed that nearly all the cars on the road in 1990 will have been manufactured with the same emission control technology and that the maintenance and fuel penalties are the same for all years of a car's life. These assumptions understate the cost penalties for older cars in 1990 since the penalties may increase as the car is driven and since control technology may improve over the decade. Constant operating cost penalties are assumed because there is no data to estimate a more complex pattern and because this study prefers to err on the side of udnerstating the burden to owners of older cars.

Table 4-4 presents estimates of national annual operating costs by income group in 1980 and 1990. Both the dollar costs and the relative cost burdens are listed. Table 4-4 reveals the same distributional pattern that was evident for the annual ownership cost burdens: the dollar operating cost burdens increase with income, while the percentage of income represented by these costs

Table 4-4. National Annual Operating Costs by Income Group in 1980 and 1990

Income Group	1980		1990	
	Costs ($)	Relative Costs (%)	Costs ($)	Relative Costs (%)
<3	20.23	1.12	16.03	0.89
3-5	37.48	0.94	24.64	0.62
5-7	48.56	0.81	30.58	0.51
7-10	59.48	0.70	36.59	0.43
10-15	84.32	0.67	46.17	0.37
15-25	106.05	0.53	55.67	0.28
25+	132.47	0.44	62.96	0.21
Inequality Ratio		2.5		4.2

decreases with income. Thus, the operating cost burdens of automobile emission control are also distributed in a regressive manner.

While the general distributional patterns are similar for ownership and operating cost burdens, the magnitudes are quite different. Operating cost burdens are both larger and distributed in a more regressive manner than the ownership cost burdens. While the national ownership burdens in 1990 range from $7.59 to $46.68, depending on the income group, the operating burdens range from $16.03 to $62.96. The national inequality ratio, used as a measure of regressivity, is 2.6 for ownership costs and 4.2 for operating costs in 1990.

Table 4–4 also indicates that the pattern of operating costs is more regressive in 1990 than in 1980, a trend which was evident for ownership costs. Although annual operating cost burdens become smaller as the maintenance and repair costs and fuel economy penalties are reduced after the first few years of stringent emission controls, the burden tends to become more heavily concentrated in the lower income groups.

STOCKHOLDER AND TAXPAYER COSTS

Automobile company stockholders and federal taxpayers will bear part of the burden of automobile emission control costs because automobile company profits will be lower than they would be without the emission control program. Profits will be lower because the new car price increases will be smaller than the per car cost incurred by automobile companies for controls. Since the corporate tax rate is 48 percent, approximately one-half of the decline in corporate profits will be borne by federal taxpayers in the form of diminished tax revenues.

The magnitudes of the pretax corporate profit declines in 1980 and 1990 depend upon the per car emission control costs, the new car price increases, and the number of new car sales for 1980 and 1990 model year cars. This formula is given in Equation 4.7.

$$A_t = (C^t - P^t) N^t \tag{4.7}$$

where: A = change in pretax profits of automobile manufacturers

C = per car emission control costs

P = new car price increase

N = total new car sales by domestic automobile manufacturers

t = analysis year

Per car emission control costs and new car price increases for 1980 and 1990 model year cars were estimated in Chapter Three. These estimates result in per car profit reductions of $26.50 in 1980 and $17.60 in 1990. It is assumed that ten million cars will be sold by domestic manufacturers in 1980 and 1990. (In 1973, 9,677,000 cars were sold by domestic manufacturers).[2]

The resulting estimates of declines in corporate profits and federal corporate tax revenues in 1980 and 1990 are given below.

Year	Pretax Profit decline	Posttax profit decline	Corporate tax decline
1980	$265,000,000	$132,500,000	$132,500,000
1990	$176,000,000	$ 88,000,000	$ 88,000,000

The allocation of the corporate profits burden and taxpayer burden among households in various income groups was done using the following formulas:

stockholder cost by income group

$$S_i = \frac{s_i \cdot .5\,A}{F_i} \tag{4.8}$$

taxpayer cost by income group

$$T_i = \frac{t_i \cdot .5\,A}{F_i} \tag{4.9}$$

where: S = stockholder burden

s = fraction of total corporate dividends received

A = pretax change in automobile company profits

F = number of households

T = taxpayer burden

t = fraction of total federal taxes paid

i = income group

The estimates of the annual stockholder and taxpayer cost burdens by income group for 1980 and 1990 are listed in Table 4-5.[3] The average burdens vary considerably by income group in both years. For 1990, the burden varies from $.36 for households in the lowest income group to $21.00 for

Table 4-5. National Annual Stockholder and Taxpayer Burdens by
Income Group in 1980 and 1990

Income Group	Stockholder Costs ($)	Taxpayer Costs ($)	Total Costs ($)	Total Relative Costs (%)
		1980		
<3	.28	.29	.57	.03
3-5	1.08	.76	1.84	.05
5-7	.91	1.11	2.01	.03
7-10	1.06	1.81	2.87	.03
10-15	1.91	2.82	4.73	.04
15-25	3.81	3.86	7.67	.04
25+	21.46	11.71	33.17	.11
		1990		
<3	.68	.18	.36	.02
3-5	.68	.48	1.16	.03
5-7	.58	.70	1.28	.02
7-10	.67	1.15	1.82	.02
10-15	1.21	1.79	3.00	.02
15-25	2.41	2.45	4.86	.02
25+	13.59	7.41	21.00	.07

households in the most affluent group. In 1980 the variation in dollar burdens is even greater, ranging from $.57 to $33.17. As expected, the dollar burden of these stockholder and taxpayer costs fall more heavily on higher income groups.

The percentage of income taken is roughly the same for all but the highest income groups. The relative burdens are listed in the last column of Table 4-5. Except for households in the highest income group, the stockholder burden is about .04 percent of household income in 1980 and .02 percent of household income in 1990. The percentage burden is about three times as great for households in the highest income group.

There are several assumptions implicit in the procedures and data used to estimate stockholder and taxpayer cost burdens. Both estimates are based on a moderate decline in pretax automobile company profits in 1980 and 1990 from what they would be without emission control costs. There is a great deal of uncertainty about these estimates since the whole milieu of the automobile market may be very different in 1980 and 1990 from what it is today. As discussed in Chapter Three, the automobile companies may actually increase their profits as a result of emission control related costs. However, whether an increase or a decrease, these profit changes amount to a small fraction of total emission control costs.

The calculation of stockholder burden by income group also assumes that the income profile of automobile company stockholders in 1980 and 1990 is the same as the income profile of those receiving all capital income in 1970 and that automobile company stockholders are not concentrated in particular SMSAs or areas of the country. If wealth becomes distributed in a less concentrated manner in the future, the burdens borne by lower income groups will be greater than estimated here. Thus, the estimates here may understate the actual burdens borne by lower income households in 1980 and 1990.

The taxpayer burden calculation assumes that reduced corporate tax revenues will be compensated for by increasing all federal taxes proportionally. However, several other alterations in federal government tax and expenditure policy may result from decreased profits tax revenue, and such alternatives have very different distributional consequences. For example, the government could either increase the gift tax rate to obtain more revenue or decrease federal welfare assistance to decrease expenditures by the same amount. The ultimate incidence of any of these tax or expenditure policies would be difficult to determine.

NOTES TO CHAPTER FOUR

1. Internal Revenue Service, *Statistics of Income – 1970, Individual Income Tax Return* (Washington, D. C.: U.S. Government Printing Office, 1972).
2. Motor Vehicle Manufacturers' Association, *1973/74 Automobile Facts and Figures,* (Detroit, Michigan: Motor Vehicle Manufacturers' Association, 1974), p. 6.
3. Data on the fraction of total corporate dividends by income group and the fraction of total federal taxes by income group were obtained from R.A. Musgrave et al., "The Distribution of Fiscal Burdens and Benefits", (Harvard Institute of Economic Research Discussion Paper Number 319, Harvard University, Cambridge, Massachusetts, September 1973), pp. 5, 13.

Chapter Five

Estimation of Air Quality Benefits

Benefits of the auto emission control plan are measured in this study as improvements in the concentrations of the three automobile related pollutants: carbon monoxide (CO), nitrogen oxides (NO_x), and photochemical oxidants (O_x). The first two pollutants are emitted directly by automobiles; the last, O_x, is produced by atmospheric reactions involving NO_x and hydrocarbons (HC), also emitted by automobiles.

The overview in Chapter Two identified four successive steps to estimate air quality benefits: (1) determine auto emission rates for cars on the road with and without controls; (2) calculate total auto emissions in each urban area under the controlled and uncontrolled cases; (3) estimate pollutant concentrations under these two circumstances; and (4) calculate air quality benefits from controls for each geographic area and, using these geographic figures, calculate benefits for each income group.

This chapter provides a detailed description of these procedures and the data needed along the way. Like its equivalent in Chapter Four, the discussion of benefit procedures and data is somewhat technical, particularly where some statistical estimation procedures are presented. But this discussion is necessary to document the estimation procedures and the data on which the results presented in Part II of this study depend.

AUTOMOBILE EMISSION RATES

Emission rates of CO, HC, and NO_x from cars depend on a great many factors.[1] Among the most important influences on emissions per mile as the car is driven are the engine type, the emission control technology, and the way the car is being operated (accelerating, decelerating, cruising, or idling). Emissions at start-up when the engine is cold are greater than those from a "hot start" when the engine is warmed up in advance. To make matters more complicated, reported emission rates vary with the method used to measure emissions.

To standardize emission measurements and determine compliance with emission control standards, the federal EPA has developed test procedures to measure car emissions. These procedures are designed to measure exhaust emissions during typical driving conditions in urban areas. All emission rates in this study are measured in terms of the most recent test procedure–the 1975 Constant Volume Sampling Procedure (CVS-CH). This test procedure is based on a 23-mile driving cycle designed to simulate average driving in downtown Los Angeles.[2]

Tests on new cars in each model year using this procedure generate estimates of the average emissions per mile for cars in the standardized driving cycle. More information is needed to estimate emission rates for cars on the road in 1980 and 1990. Emission rates increase as a car is driven, because emission controls deteriorate, and also vary with the average operating conditions, as mentioned above.

The EPA has developed a formula and supporting data which allows emissions per mile to be estimated for cars on the road in 1980 and 1990. The formula takes into account the model year of the car, its age, and the average speed that it is driven. Average speed is used to proxy the various influences of acceleration, deceleration, and other operating conditions on car emission rates. The formula for calculating the emissions per mile for a car in a given age group is the following:

$$e_{jt} = c_j \cdot d_{jt} \cdot s_j \qquad\qquad (5.1)$$

where: e = exhaust emissions per mile

c = original new car emission rate (measured by CVS-CH procedure)

d = deterioration factor

s = speed correction factor

j = car age group

t = analysis year

New Car Emission Rates

The EPA new car exhaust emission rates are listed in Table 5-1 and presented graphically in Figure 5-1.[3] Because the procedure to certify compliance with the new car emission standards now requires that each car tested meet the standard, rather than that the average not exceed the standard, these average values are somewhat less than the standards in each year.

Table 5-1. Average New Car Exhaust Emissions by Model Year

Model Year	Average Exhaust Emissions (g/mile)		
	CO	HC	NO_x
1970	36	3.6	5.1
1971	34	2.9	4.8
1972	19	2.7	4.8
1973	19	2.7	2.3
1974	19	2.7	2.3
1975	12.5	1.3	2.2
1976	1.8	0.23	1.6
1977 and thereafter	1.8	0.23	0.31

SOURCE: U.S. Environmental Protection Agency, *Supplement Number 2 for Compilation of Air Pollutant Emission Factors,* Report AP-42, 2nd ed. (Washington, D.C.: U.S. Government Printing Office, September 1973), p. 3.1.2-2.

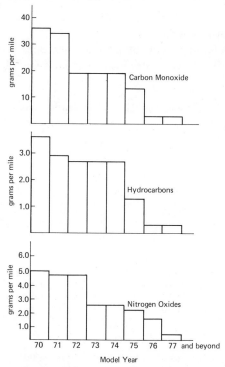

SOURCE: U.S. Environmental Protection Agency, *Supplement Number 2 for Compilation of Air Pollutant Emission Factors,* Report AP-42, 2nd ed. (Washington, D.C.: U.S. Government Printing Office, September 1973), p. 3.1.2-2.

Figure 5-1. Average New Car Exhaust Emissions for Model Years 1970 to 1977 and Beyond

Both CO and NO_x are emitted only from the engine exhaust. Calculating HC emissions involves an additional complication since HC are emitted from crankcase and fuel tank evaporation as well as from the exhaust. The EPA estimated that in 1970 cars, evaporative emissions of HC were 3.0 grams per mile.[4] Beginning in 1971 cars, a fuel evaporation control system was added to control HC evaporative losses to 0.5 grams per mile in 1971 and 0.2 grams per mile in 1972 and thereafter. The figures in this study incorporate these estimated evaporative losses.[5]

Emission Deterioration Factors

Emission rates increase as cars are driven because control parts wear out, the engines are less likely to be tuned properly, and the effectiveness of emission control technology deteriorates. While the precise increase for a particular car depends on how many miles and under what conditions it is driven, the EPA has estimated average deterioration factors.[6] These EPA figures are listed in Table 5-2.

Speed Correction Factors

The relationship between pollutant emission and average speed is complicated and differs for the three pollutants. Emissions of both CO and HC

Table 5-2. Exhaust Emission Deterioration Factors for CO, HC, and NO_x

Model Year	0	1	2	Vehicle Age 3	4	5	6	7	8	9+
				CO						
1970–1974	1.00	1.18	1.32	1.38	1.40	1.44	1.47	1.50	1.51	1.56
1975	1.00	1.04	1.30	1.36	1.43	1.44	1.49	1.56	1.63	1.69
1976+	1.00	1.16	1.34	1.50	1.62	1.75	1.88	2.00	2.10	2.22
				HC						
1970–1974	1.00	1.05	1.10	1.13	1.15	1.17	1.20	1.22	1.24	1.26
1975	1.00	1.00	1.13	1.22	1.29	1.37	1.43	1.50	1.56	1.63
1976+	1.00	1.14	1.30	1.44	1.55	1.67	1.77	1.88	1.96	2.07
				NO_x						
pre 1973	1.00	1.00	1.00	1.00	1.00	1.00	1.00	1.00	1.00	1.00
1973–1974	1.00	1.11	1.18	1.20	1.21	1.22	1.23	1.24	1.25	1.26
1975	1.00	1.00	1.18	1.23	1.23	1.41	1.45	1.45	1.45	1.45
1976	1.00	1.02	1.07	1.10	1.13	1.17	1.19	1.21	1.24	1.26
1977+	1.00	1.17	1.37	1.53	1.67	1.82	1.94	2.06	2.17	2.32

SOURCE: U.S. Environmental Protection Agency, *Supplement Number 2 for Compilation of Air Pollutant Emission Factors,* Report AP–42, 2nd ed. (Washington, D.C.: U.S. Government Printing Office, September 1973), p. 3.1.2–6.

are *higher* at lower speeds and during stop and start driving, when engine temperatures are low. These same low speed, low engine temperature conditions result in *lower* than normal emissions of NO_x. The EPA has developed empirical estimates of the relationships between average speed and emission rates, expressed as speed correction factors relative to the CVS-CH test cycle average of 19.7 miles per hour.[7] The EPA speed correction factors for the three pollutants are graphed in Figure 5-2.

Emissions Per Mile by Car Age
These EPA data permit the calculation of 1980 and 1990 emission rates for the six car age categories used in the estimation procedure. Table 5-3 presents these estimates separately for central city and suburban areas. In applying the speed correction factors, average speeds are assumed to be 15 miles per hour in central city areas and 30 miles per hour in suburban areas. The 1980 and 1990 figures in Table 5-3 represent the controlled situation, in which the

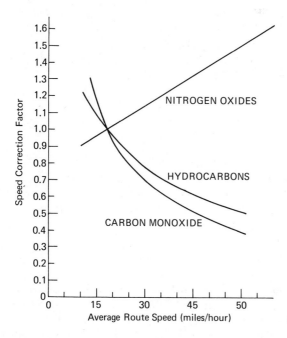

SOURCE: U.S. Environmental Protection Agency, *Supplement Number 2 for Compilation of Air Pollutant Emission Factors,* Report AP–42, 2nd ed. (Washington, D.C.: U.S. Government Printing Office, September 1973), p. 3.1.1-7.

Figure 5-2. Average Speed Correction Factors for All Model Years

Table 5-3.   Air Pollutant Emission Rates by Car Age Category for Central City and Suburban Areas

Car Age	Uncontrolled CC	Uncontrolled SUB	1980 CC	1980 SUB	1990 CC	1990 SUB
			CO (grams/mile)			
New	44.3	24.5	2.21	1.22	2.21	1.22
1	52.3	28.9	2.57	1.42	2.57	1.42
2	58.5	32.3	3.00	1.64	3.00	1.64
3-4	61.6	34.0	3.45	1.91	3.45	1.91
5-7	65.1	36.0	30.74	17.00	4.16	2.30
8+	68.2	37.7	69.08	38.19	4.92	2.72
			HC (grams/mile)			
New	7.3	5.8	.47	.38	.47	.38
1	7.5	5.9	.51	.40	.51	.40
2	7.7	6.1	.55	.43	.55	.43
3-4	7.8	6.2	.61	.47	.61	.47
5-7	8.1	6.3	3.22	2.23	.68	.52
8+	8.4	6.5	5.55	3.69	.76	.57
			NO_x (grams/mile)			
New	4.9	5.8	.30	.35	.30	.35
1	4.9	5.8	.35	.41	.35	.41
2	4.9	5.8	.40	.48	.40	.48
3-4	4.9	5.8	1.20	1.43	.47	.56
5-7	4.9	5.8	2.54	3.02	.57	.68
8+	4.9	5.8	4.85	5.76	.68	.81

new car emission rates in Table 5-1 apply. The first column lists the uncontrolled case in which all cars are manufactured with the 1970 emissions characteristics. The figures in Table 5-3 for controlled and uncontrolled emission rates constitute the building blocks for this study's estimates of air quality benefits from the current emission control program.

URBAN AREA AUTOMOBILE EMISSIONS

The total weight of pollutants emitted from automobiles in each geographic area is calculated from these per mile emissions figures and from estimates of the total vehicle miles of travel (VMT) driven by car age. Total emissions are the sum of emissions from each car age group, and are calculated according to Equation 5.2.

$$E_t = \sum_{j=1}^{6} (VMT_{jt} \cdot e_{jt}) \tag{5.2}$$

where E = total automobile emissions

 VMT = total vehicle miles of travel

    e = emissions per mile

    j = car age group
    t = analysis year

The VMT data used in Equation 5.2 were estimated by calculating the average miles traveled by car age for households in various income groups and then aggregating over all income groups, i.e., Equation 5.3.

$$VMT_{jt} = \sum_{i=1}^{7} (V_{ijt} \cdot N_{it}) \qquad (5.3)$$

where: VMT = total vehicle miles of travel

     V = average annual vehicle miles of travel

     N = number of households

     j = car age group

     i = income group

     t = analysis year

These travel data were explained in Chapter Two.

The calculation of total automobile emissions in this study assumes that households living in the central city drive only in the central city and that those in the suburbs drive only in the suburbs. Such an assumption is inaccurate, since most cars are driven outside the urban area as well as within both central city and suburban areas. Since no information is available on the average mileage drivin by cars of various ages in central city and suburban areas, it is impossible to relax this assumption. However, the bias in total emission for the two subareas is probably small. Trips outside the urban area account for a small fraction of total miles traveled for most cars.[8] Nonwork trips—trips for shopping, school, and other family business—are usually made to destinations close to home. Work trips may often involve commuting between suburban and central city areas, but most the trip is within one of these subareas. In the most familiar case, the suburban commuter travels much of his commute on suburban highways even if his destination is in the central city.

URBAN AREA AIR POLLUTANT CONCENTRATIONS

The technique used to estimate 1980 and 1990 air pollutant concentrations for urban areas in the proportional rollback method, in which concentration is assumed proportional to total emissions. The rollback formula used in this study has the following form:

$$C_t = \left[\frac{E_t + (r \cdot E_b)}{E_b + (r \cdot E_b)} \right] C_b \tag{5.4}$$

where: C = concentration

E = total automobile emissions

r = nonautomobile emissions/automobile emissions

t = analysis year

b = base year

The denominator in the bracketed portion of Equation 5.4 is total base year emissions and the numerator is total analysis year emissions. Thus estimating concentrations in 1980 and 1990 requires information on base year emissions, analysis year emissions, and base year concentrations.

Base Year Emissions

The base year in this study is 1970. Automobile emission in 1970, E_b, were estimated using the procedures described in the last section. Estimates of the ratio of nonautomobile emissions to automobile emissions, r in Equation 5.4, are national averages.[a] While this ratio differs among urban areas, it was not possible to obtain separate estimates for the 487 geographic areas in this study.

Analysis Year Emissions

Several simplifying assumptions were made to calculate emissions in 1980 and 1990 under controlled and uncontrolled cases. For one thing, non-automobile emissions are assumed to be at the same levels in 1980 and 1990 as they were in 1970. Without information about nonauto emissions in future years, it is logical to assume a constant level of such emissions. The calculation of auto emissions in 1980 and 1990 also assumes that the total number of auto miles traveled is the same in 1980 and 1990 as in 1970. Only the emissions characteristics of cars are assumed to change. Changes in these assumptions would of course alter the calculation of analysis year emissions. Experiments with different assumptions (not reported here) showed that the *distributional* conclusions

---

[a]The following values for r were used:[9]

CO : 0.163
HC : 0.351
NO$_x$: 0.961

of this study are not very sensitive to these assumptions, although the estimates of overall national benefits are influenced by these assumptions.

Base Year Concentrations

The major difficulty in calculating 1980 and 1990 concentrations using the rollback formula is measuring 1970 concentrations. Data on 1970 concentrations of CO, NO_x and O_x are only available for a limited number of central cities. To estimate 1970 concentrations in the other central cities and in suburban areas, relationships between pollutant concentrations and automobile emissions per square mile were estimated for the 16 central cities with the requisite pollution data. For photochemical oxidants, a relationship between oxidant concentration and hydrocarbon emissions per square mile was estimated. These relationships were estimated under the assumption that pollutant concentration is an exponential function of *auto* emission density. An exponential relationship is assumed because nonautomobile densities are likely to be higher where automobile densities are higher. These equations should have positive intercepts because of the nonautomobile contributions.

The estimated relationships between 1970 concentrations and auto emissions densities are given below. The *t* statistics of the emissions density coefficients are in parentheses.

base year CO concentration

$$LC = .6339 + 1.345\,CE \quad R^2 = .67 \tag{5.5}$$
$$(5.3)$$

base year O_x concentration

$$LO = .3085 + 11.16\,HE \quad R^2 = .55 \tag{5.6}$$
$$(4.1)$$

base year NO_x concentration

$$LN = 3.173 + 6.993\,NE \quad R^2 = .16 \tag{5.7}$$
$$(1.6)$$

where: LC = natural log of CO concentration (8 hour annual maximum in ppm)

LO = natural log of O_x concentration (1 hour annual maximum in 100 x ppm)

LN = natural log of NO_x concentration (24 hour annual average in 1000 x ppm)

CE = CO automobile emissions per square mile

HE = HC automobile emissions per square mile

NE = NO_x automobile emissions per square mile

These equations were used to estimate 1970 baseline concentrations for the central city and suburban portions of the 243 SMSAs.[b] The 1970 concentrations for the non-SMSA area are assumed to equal the equation intercepts, since average automobile emissions densities in those areas are very close to zero.

AIR QUALITY BENEFITS

The previous data and procedures are all designed to yield estimates of the air quality benefits Americans will experience in 1980 and 1990 because of federal automobile emission controls. This section presents the final steps to calculate air quality benefits for each geographic area, for the nation as a whole, and for various income groups.

Benefits by Geographic Area

Air pollutant concentrations in 1980 and 1990 were calculated for two levels of automobile emissions: (1) the uncontrolled case in which all vehicles manufactured after 1970 have the 1970 emissions characteristics; and (2) the controlled case in which the current auto emission standards mandated by Congress in the 1970 Clean Air Act and administrively set by the EPA apply. The difference between these two sets of pollutant concentrations measures the benefits accruing to households in the area from the current program. The benefit for each pollutant can be expressed algebraically as:

$$B = C^u - C^c \qquad (5.8)$$

where: B = air quality benefit

C^u = uncontrolled concentration

C^c = controlled concentration

Air quality benefits were calculated using this formula for all 487 geographic

---

[b]These estimated equations conform rather well to prior expectations. All have positive intercepts which provide plausible estimates of the concentrations which would prevail where automobile emission densities are very low. As the R^2 and t statistics indicate, concentrations of CO and O_x are predicted well by these equations. The coefficients of CO and HC are significant at the .01 level. The estimated relationship for NO_x is less satisfactory, since the R^2 is low and the coefficient only statistically significant at the .10 level. However, past NO_x concentration measurements have been found to be unreliable, and the EPA is now devising another measuring procedure. The poor explanatory power of the estimated relationship may be due to the inaccuracy of the NO_x concentration measures.

areas (243 central cities, 243 suburban areas, and the residual non-SMSA area).

Total National Benefits

Total national benefits should measure the reduced exposure of *all* Americans to the three automobile air pollutants. Using the 487 geographic-specific benefit estimates and estimates of the number of households in each of these areas, the total national reduction in households' exposure to these three pollutants can be calculated. The formulas for calculating the total national benefit and the average per household benefit for each pollutant are the following:

total national benefit

$$NB = \sum_{m=1}^{487} (B_m \cdot N_m) \tag{5.9}$$

average national benefit

$$AB = \frac{NB}{\sum_{m=1}^{487} N_m} \tag{5.10}$$

where: NB = total national benefit of automobile emission control (ppm-households)

AB = average benefit of automobile emission control (ppm)

B = benefit

N = number of households

m = geographic area

Benefits by Income Group

The average air quality benefit to a household in the ith income group is a weighted sum of the improvements in all areas, the weights being the fractions of households in the ith income group living in each subarea.[10]

$$B_i = \sum_{m=1}^{487} (B_m \cdot f_{im}) \tag{5.11}$$

where: B = average air quality

f = fraction of households

i = income group

m = geographic area

The calculation in Equation 5.11 applies to households throughout the United States. A completely analogous formula is used to calculate the average benefit by income group for the households in each SMSA as a weighted sum of the central city and suburban improvements. In this formula, the weights are the fractions of households in the ith income group living in each subarea.

$$B_i = \left(B^c \cdot f_i^c\right) + \left(B^s \cdot f_i^s\right) \tag{5.12}$$

where: B = average air quality benefit

f = fraction of households

i = income group

c = central city

s = suburb

$$(f_i^c + f_i^s = 1)$$

All of the benefit calculations in this study make two implicit assumptions which should be mentioned. For one thing, the calculation of benefits assumes that households experience air quality benefits only where they live. Of course household memebers spend some time outside their residence area and therefore receive some benefits from breathing cleaner air in a wide variety of areas. It would be desirable to estimate the time various household members spend outside their home area and to calculate a more precise measure of benefits, but the data for this task does not exist. However, most household benefits will be derived from nearby improvements. Nonworking members spend most of their time near home, and many workers work near where they live. Even for commuters, a large part of the day will be spent either driving through or living in their home area.

The benefit estimates also ignore the possibility that for renters some of the air quality benefits may be offset by increases in their rents due to the improved local air quality. If rents did increase, some of the benefits of auto emission control would accrue to landlords rather than to residents of the area. It is hard to gauge the quantitative magnitude of these price adjustments in the rental housing market and the household relocations which may accompany them. A recent study of the influence of auto air pollutant concentrations on rents in Boston and Los Angeles suggests that some price adjustments of this sort may occur.[11] But estimating these complicated housing market adjustments for the geographic areas considered here was not attempted in this study.

NOTES TO CHAPTER FIVE

1. U.S. Department of Health, Education and Welfare, *Control Techniques for*

Carbon Monoxide, Nitrogen Oxide, and Hydrocarbon Emission from Mobile Sources (Washington, D.C.: U.S. Government Printing Office, March 1970), pp. 1–11 to 1–17.

2. A description of the CVS−CH federal test procedure and a discussion of the various alternative test procedures are found in the following: J.B. Heywood, "Impact of Emission Controls: 1968–1974," in F.P. Grad et al., *The Automobile and the Regulation of Its Impact on the Environment* (Norman, Oklahoma: University of Oklahoma Press, 1975), pp. 115–123.

3. U.S. Environmental Protection Agency, *Supplement Number 2 for Compilation of Air Pollutant Emission Factors,* Report AP–42, 2nd ed. (Washington, D.C.: U.S. Government Printing Office, September 1973), p. 2.1.2.–2.

4. Ibid., p. 3.1.2–4.

5. There is some evidence that the post-1971 evaporative emission controls were not as successful as the EPA data indicate. See Heywood, "Impact of Emission Controls: 1960–1974," pp. 130–132.

6. U.S. Environmental Protection Agency, *Emission Factors,* p. 3.2.3–6.

7. Ibid., p. 3.1.1–7.

8. U.S. Department of Transportation, *Household Travel in the United States,* Report No. 7 of the Nationwide Personal Transportation Study (Washington, D.C.: U.S. Department of Transportation, December 1972), p. 4.

9. These values were obtained from the following source: U.S. Environmental Protection Agency, *The National Air Monitoring Program: Air Quality and Emissions Trends,* Annual Report, vol. 1 (Research Triangle Park, North Carolina: U.S. Environmental Protection Agency, August 1973), p. 1–7.

10. The household fractions are obtained from the following: U.S. Bureau of the Census, *Census of Housing: 1970 Metropolitan Housing Characteristics,* Final Report HC(2)–1 to HC(2)–244 (Washington, D.C.: U.S. Government Printing Office, 1972).

11. This study was conducted by the author and Robert MacDonald and is reported in, National Academy of Sciences, *A Report by the Coordinating Committee on Air Quality Studies, Volume IV: The Costs and Benefits of Automobile Emission Control,* (Washington, D.C.: U.S. Government Printing Office, September 1, 1974), pp. 221–242.

Part II

The Results

Distribution of Emission Control Costs

This chapter begins Part II of the study, the presentation of empirical results. Discussion of the procedures and the data in Part I of the study provided the skeleton and some of the vital organs for these empirical results. This chapter and the next provide the flesh to fill out the body for the current federal auto emission control program. Chapter Six considers the cost impacts and Chapter Seven considers the benefit impacts. Chapter Eight presents the results for several plausible alternative schemes, and Chapter Nine summarizes the major conclusions of the study.

Both Chapters Six and Seven begin by presenting the national impacts of the current program. These national impacts are primarily designed to provide a benchmark for average costs and air quality benefits, since this study emphasizes *variations* in impacts for different household groups in the United States. The other sections of the two chapters give detailed results for income groups, urban areas, population size groups, and regions of the country.

TOTAL NATIONAL COSTS

Improving air quality by imposing stringent emission standards on all automobiles is clearly a costly program. Table 6-1 presents the national annual costs of the automobile emission control program in 1980 and 1990. In 1980, the total annual cost is almost $6 billion, which is $94.10 per household. In 1990, when the annual costs for all cars of meeting the stringent emission standards will be at a steady state minimum, the total cost of emission control is nearly $4 billion, or $62.50 per household.

Figure 6-1 shows a breakdown of costs by major category in 1980 and 1990. Operating costs are the most important category of total costs, particularly in 1980. In 1980, operating costs account for over two-thirds

Table 6-1.   National Annual Costs of Emission Control by Cost
Component in 1980 and 1990

| | Control Costs ($ millions) | |
Cost Component	1980	1990
Ownership	1,678	1,463
Operating	4,009	2,316
Stockholder and Taxpayer	283	180
Total	5,970	3,959
Average cost per household	$94.10	$62.40

of the control costs. The decline in operating costs from over $4 billion in 1980
to $2.3 billion in 1990 is largely responsible for the decline in total control
costs. In contrast, ownership costs only decline from $1.7 billion in 1980 to
$1.5 billion in 1990. Stockholder and taxpayer costs remain a small part of the
total cost burden in both years.

These estimates are probably conservative estimates of the true
annual control costs in 1980 and 1990. For one thing, these totals are based on
the number of households in 1970. If the number of households is greater in
1980 and 1990 and if households in these future years own cars and travel at
the same rates as households in 1970, total control costs will be greater than
predicted. Control cost totals would also be greater if the costs for enforcement
and administration of the control program were included and if the price of

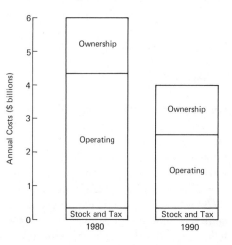

Figure 6-1. National Annual Costs of Emission Control by Cost
Component in 1980 and 1990

gasoline in 1980 and 1990 is higher than predicted here ($.50 per gallon in 1972 dollars).

There are also factors which may result in lower control costs in 1980 and 1990. If households in 1980 and 1990 buy fewer cars and drive them fewer miles per year on average than households in 1970, the total costs of emission control might be less than calculated using 1970 car ownership and vehicle miles of travel estimates. Also, the cost penalties for cars to meet the standards under the current emission control program may be lower than estimated here, particularly for cars manufactured in the 1980s. The cost figures used here are based on current estimates of the costs of likely emission control technology. Improved technology or declining production costs for the same technology may results in smaller initial costs and smaller gas and maintenance penalties for cars manufactured in the 1980s. However, it is doubtful that the costs of meeting the stringent emission control standards in the 1980s will be substantially lower than estimated in this study, since these estimates rest on optimistic views of the pace of technological progress and the prospects of alternative engine technolgies.

VARIATION BY INCOME GROUP

One of the major objectives of this study is to determine the variation in costs for households in different income groups. This section takes a national perspective, inquiring whether there are systematic differences in control costs by income group for the country as a whole. Table 6–2 presents the dollar costs borne by households in different income groups in 1980 and 1990. Dollar costs range from $27.56 for households with less than $3,000 per year to $225.75 for households earning more than $25,000 per year.

Table 6–2 also lists the relative burden of emission control costs, defined as the percentage of household income accounted for by automobile

Table 6–2. National Annual Costs and Relative Cost Burdens by Income Group in 1980 and 1990

Income Group	1980 Costs ($)	1980 Relative Costs (%)	1990 Costs ($)	1990 Relative Costs (%)
<3	27.56	1.5	23.98	1.3
3–5	54.07	1.4	39.93	1.0
5–7	70.60	1.2	50.25	0.8
7–10	86.37	1.0	60.39	0.7
10–15	125.35	0.8	79.79	0.6
15–25	159.75	0.8	98.38	0.5
25+	225.75	0.8	130.63	0.4
Inequality Ratio		2.0		3.1

emission control costs. These figures indicate that the program will have disturbing equity implications since control costs will be more burdensome to
lower income groups. While the dollar costs increase with income, the percentage
of household income accounted for by emission control costs declines steadily
with income. This *regressive* cost pattern (to use economic terminology) is
illustrated graphically in Figure 6-2. The inequality ratio, defined as the relative burden for the under $3,000 group divided by the relative burden for the
over $25,000 group, also indicates the greater burden on lower income groups.
Inequality ratios are 2.0 in 1980 and 3.1 in 1990.

Such results are particularly worrisome since this study has attempted to understate the burdens falling on lower income households. Cost burdens
may fall even more heavily on low income households because the cost penalties
for owners of older cars may be greater than the estimates in this study. The results assume that maintenance and fuel penalties are constant over a car's life.
In reality, these penalties are greater during the later years of a car's life. In
addition, possible industry strategies to maintain profits, such as by increasing
replacement parts prices, were not taken into account. If the automobile companies do undertake usch a strategy, some of the cost burden will shift from
high income stockholders and federal taxpayers to low income owners of used

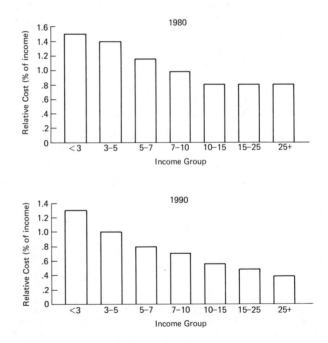

Figure 6-2. National Relative Cost Burdens by Income Group
in 1980 and 1990

cars. For these reasons, the distributional patterns estimated in this study probably understate the regressivity of automobile emission control cost burdens.

Table 6-2 and Figure 6-2 also reveal a disturbing change in the income distributional pattern over time. While average emission control costs are lower in 1990 than in 1980, they fall more heavily on the poor in the later year. This change is shown in the inequality ratio, which increases from 2.0 in 1980 to 3.1 in 1990. Thus, in 1990 the percentage of income paid by a household in the lowest income group for auto emission controls will be over three times the percentage paid by a household in the most affluent group.

VARIATION BY URBAN AREA

While the national estimates provide the best summary of the overall impacts of the control strategy, they ignore the variation in the cost pattern among geographical areas. Cost estimates were prepared for 487 geographical subareas—the central city and suburban portions of 243 SMSAs and one residual non-SMSA area—in 1980 and 1990. Discussing all these estimates would be impractical and unnecessary. Five SMSAs were selected that illustrate the diversity of cost pattern in the whole sample. These five SMSAs are Birmingham, Boston, Los Angeles, St. Louis, and Topeka. No 1980 estimates are presented in the rest of this chapter because the comparison of national cost impacts illustrates the two essential points there—that annual costs are both lower and distributed more regressively in 1990 than they are in 1980.

The cost variations among SMSAs are evaluated along two dimensions—the average level of control costs, and the distribution of control costs by income group.

Average Cost

Table 6-3 presents the 1990 average dollar costs borne by households in the central city and suburban portions of the five selected SMSAs. These cost figures are depicted graphically in Fig. 6-3. In this group of ten geographic areas, average annual costs per household range from $33.88 for the city of Boston to $77.34 for the suburbs of Topeka. Thus, one conclusion is clear—there are wide variations in average control costs among geographic areas. Moreover, these variations appear to result from several systematic trends.

Costs are uniformally higher in suburban rather than central city areas, a trend which reflects two important differences in car ownership patterns. The income profile of suburban residents is more skewed to high income households, who own more and newer cars. In addition, car ownership levels are higher in the suburbs when households in the same income group are compared. Households choosing to live in the suburbs trade off greater transportation expenses, primarily automobile ownership and operation costs, against a

Table 6–3.   Average Annual Costs for Five SMSAs in 1990

SMSA	*Dollar Costs ($)*
Birmingham	
CC	54.44
SUB	70.23
Boston	
CC	33.88
SUB	65.28
Los Angeles	
CC	62.68
SUB	74.26
St. Louis	
CC	39.79
SUB	72.07
Topeka	
CC	67.40
SUB	77.34

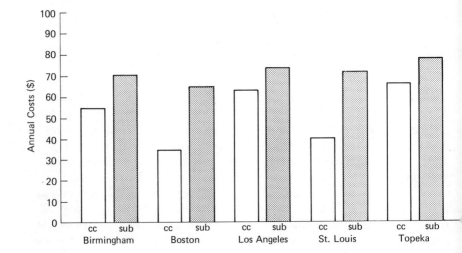

Figure 6–3.   Average Annual Costs for Five SMSAs in 1990

reduction in housing prices because land is less expensive in areas less accessible to major employment areas. The sum of housing costs and transportation costs may be roughly the same for central city and suburban households in the same income group. But households who choose to live in suburban areas will bear considerably greater automobile emission control costs.

Table 6–4 shows the differences in average control costs among central city and suburban residents in the same income group. Control costs are greater for suburban residents in all cases except the highest income group in Topeka. The difference is particularly striking for lower income households. For example, in St. Louis the cost burden for suburban households in the lowest income group ($25.47) is more than two times the burden for central city residents in the same income group ($12.04). In comparison, the burden for suburban residents in the highest income group is only 17 percent higher than that for central city residents.

The results in Tables 6–3 and 6–4 also suggest another trend in average control costs—that control costs are larger in smaller urban areas. In smaller urban areas, public transportation is poor and origins and destinations are relatively far apart because densities are low. Automobile ownership and use in these smaller areas, therefore, is more a necessity than a luxury. By contrast, in larger urban areas public transportation is often readily available and origins and destinations are closer together.

To test for differences according to population size class, the 243 SMSAs were divided into four groups based on their total population in 1970. The four categories range from SMSAs with more than four million people to

Table 6–4. Annual Dollar Costs by Income Group for Five SMSAs in 1990

SMSA	<3	3-5	5-7	Income Group 7-10	10-15	15-25	25+
Birmingham							
CC	18.44	36.01	50.69	64.60	85.98	105.41	124.89
SUB	24.34	49.27	60.57	72.63	95.13	111.97	138.80
Boston							
CC	10.16	16.57	24.95	34.81	50.28	69.61	99.34
SUB	19.91	31.69	41.17	52.08	70.88	94.10	133.03
Los Angeles							
CC	21.99	35.59	47.28	58.48	80.91	103.21	137.68
SUB	28.27	43.35	54.00	64.46	86.70	107.50	142.61
St. Louis							
CC	12.04	24.12	34.22	44.81	63.74	86.34	115.06
SUB	25.47	42.05	52.37	62.13	81.81	102.53	134.53
Topeka							
CC	24.60	44.32	56.26	65.16	89.16	106.41	136.03
SUB	37.51	50.49	59.96	74.93	89.57	112.20	126.34

SMSAs with less than 300,000 people. Table 6-5 describes these four SMSA groups and the fifth group, consisting of households living outside SMSAs. Estimates of dollar cost burdens by income group for these five population size classes are listed in Table 6-6. The variations by SMSA size class are large and conform well to expectations. Cost burdens increase as population size decreases for households in the same income group and residential location group (central city or suburb) in all cases but one. This sole

Table 6-5. Population Size Classes

SMSA size	Number of SMSAs	CC	SUB	Total	Cumulative Total	Cumulative %
			Number of Households (1000s)			
4 million +	5	6,283	4,953	11,236	11,236	17.7
1-4 million	28	5,828	8,790	14,618	25,854	40.8
300,000-1 million	68	5,089	5,615	10,704	36,558	57.7
<300,000	141	4,071	3,181	7,252	43,810	69.1
All SMSAs	242	21,271	22,539	43,810	43,810	69.1
Outside SMSA			19,586	19,586	63,396	100.0
Total U.S.		21,271	42,125	63,396	63,396	100.0

SOURCE: Author's analysis of data from U.S. Bureau of the Census, *Census of Housing: 1970 Metropolitan Housing Characteristics,* Final Report HC(2)-1 to HC(2)-244 (Washington, D.C.: U.S. Government Printing Office, 1972).

Table 6-6. Annual Dollar Costs by Income Group for Five Population Size Classes in 1990

SMSA Size	<3	3-5	5-7	7-10	10-15	15-25	25+
				Income Group			
4 million +							
CC	11.03	18.33	25.78	36.53	54.28	74.02	101.89
SUB	27.11	40.36	50.52	60.75	80.46	100.78	137.15
1-4 million							
CC	18.00	30.62	41.09	52.91	74.11	95.96	128.41
SUB	28.13	42.23	52.17	62.00	82.01	101.74	136.51
300,000-1 million							
CC	21.19	36.55	47.89	59.25	79.85	100.54	133.41
SUB	29.34	45.80	56.13	65.90	85.59	104.65	137.80
<300,000							
CC	23.29	39.79	51.47	62.05	82.80	102.69	136.10
SUB	30.12	48.62	58.34	67.40	86.86	105.25	135.65
Outside SMSA							
Total	26.74	46.25	57.06	66.00	84.38	100.06	128.55

exception is for suburban houscholds in thc highest income group, for which the burdens are roughly the same regardless of the size of the SMSA. The cost burdens for non-SMSA households seem to be about the same as the burdens for households living in the suburban areas of medium sized SMSAs (300,000-1 million persons).

The trend toward higher costs for households in smaller urban areas is much more pronounced in central city portions than in suburban areas. Figure 6-4 shows the per household costs for middle income households (7,000-$9,999) in the four SMSA size classes. Annual costs for central city households range from $36.53 to $62.05, while the range is only from $60.75 to $67.40 for suburban households.

Income Distributional Patterns

Table 6-7 presents the relative cost burdens (costs as a percentage of income) by income group and the inequality ratios for the five selected SMSAs. Control cost burdens fall more heavily on lower income groups (i.e. are regressive) in all areas. However, the degree of regressivity varies a great deal. Figure 6-5 shows the inequality ratios on a bar graph. For the five SMSAs in Table 6-7, the inequality ratio ranges from 1.7 for the city of Boston to 5.0 for the suburbs of Topeka. For the whole sample of 243 SMSAs, this ratio ranges from 1.4 for the city of Baltimore to 5.3 for the suburbs of Danbury, Connecticut.

As with average costs, these figures suggest several systematic trends in regressivity patterns among urban areas. Costs are distributed less regressively

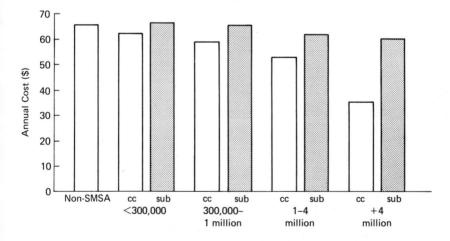

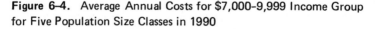

Figure 6-4. Average Annual Costs for $7,000-9,999 Income Group for Five Population Size Classes in 1990

Table 6-7. Relative Cost Burdens by Income Group for Five
SMSAs in 1990

SMSA	<3	3-5	5-7	7-10	10-15	15-25	25+	Inequality Ratio
Birmingham								
CC	1.02	.90	.85	.76	.69	.53	.42	2.4
SUB	1.32	1.23	1.01	.85	.76	.56	.46	2.9
Boston								
CC	.56	.41	.42	.40	.40	.35	.33	1.7
SUB	1.10	.79	.69	.61	.57	.47	.44	2.5
Los Angeles								
CC	1.22	.89	.79	.69	.65	.52	.46	2.7
SUB	1.57	1.08	.90	.76	.69	.54	.48	3.3
St. Louis								
CC	.67	.60	.57	.53	.51	.43	.38	1.8
SUB	1.41	1.05	.87	.73	.65	.51	.45	3.1
Topeka								
CC	1.37	1.11	.94	.77	.71	.53	.45	3.0
SUB	2.08	1.31	1.00	.88	.72	.56	.42	5.0

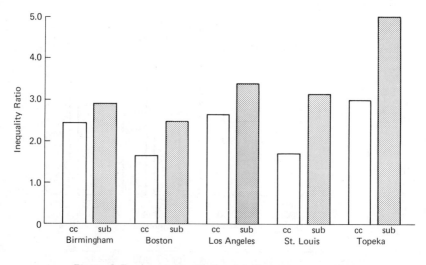

Figure 6-5. Inequality Ratios for Five SMSAs in 1990

in all central city areas than in their suburban counterparts, a trend which re-flects the lower car ownership rates among lower income households in central city rather than suburban areas. In central city areas, with their superior public transportation systems and closer origins and destinations, low income house-holds can enjoy tolerable mobility without owning a car. Low income house-holds in suburban areas are more likely to require a car to get to work, shopping areas, and other destinations. As discussed above, these suburban households may pay less for comparable housing than central city residents and thus make up for some of the added transportation expenses. But the added transportation costs involved in auto emission controls fall more heavily on these suburban poor.

Costs appear to be distributed less regressively in larger rather than smaller urban areas, although Los Angeles is an exception. A trend of this sort was expected. The public transportation facilities and close origins and destina-tions which allow poor central city households to escape the burdens of auto emission control costs are much more prevalent in larger urban areas. In smaller urban areas, low income households are more likely to pay emission control costs because they do not have good public transportation available and must use automobiles more.

These systematic trends were tested by calculating income distribu-tional patterns for the five population size classes defined above. Table 6–8 lists relative costs by income group for the central cities and suburbs of the five population size classes. The inequality ratios for each size class are given in the last column and presented in a bar graph in Figure 6–6. These results confirm the trends observed in the small sample. Inequality ratios decline consistently

Table 6-8. Relative Cost Burdens by Income Group for Five Population Size Classes in 1990

SMSA Size	<3	3-5	5-7	7-10	10-15	15-25	25+	Inequality Ratio
				Income Group				
4 million +								
CC	.61	.46	.43	.43	.43	.37	.34	1.8
SUB	1.51	1.01	.84	.71	.64	.50	.46	3.3
1-4 million								
CC	1.00	1.77	.68	.62	.57	.48	.43	2.3
SUB	1.56	1.06	.87	.73	.66	.51	.46	3.4
300,000-1 million								
CC	1.18	.91	.80	.70	.64	.50	.44	2.6
SUB	1.63	1.14	.94	.78	.68	.52	.46	3.5
<300,000								
CC	1.29	.99	.86	.73	.66	.51	.45	2.9
SUB	1.67	1.22	.97	.79	.69	.53	.45	3.7
Outside SMSA								
Total	1.49	1.16	.95	.78	.68	.50	.43	3.5

as population size increases. For central city areas, the inequality ratio varies from 1.8 in SMSAs with more than 4 million people to 2.9 in SMSAs with fewer than 300,000 persons. Suburban inequality ratios show the same trend, although the variation is less. Suburban inequality ratios range from 3.3 to 3.7. The inequality ratio for the non-SMSA category is 3.5.

A consistent picture emerges from this study of variations in cost patterns by urban area. The costs of automobile emission control are both smaller and fall less heavily on the poor in central cities compared to suburban and nonurban areas. These tendencies are much more evident in the central cities of major urban areas. Households in large central cities avoid large control costs because they tend to have relatively low auto ownership levels. These low car ownership levels in central cities reflect generally good public transportation systems and relatively close origins and destinations. Cost burdens are borne less by lower income groups because low income households are particularly likely to escape control cost burdens by not owning a car.

In contrast, households in suburban areas and in nonurban areas will bear quite large costs for the automobile emission control program. Moreover, these costs will fall much more heavily on the lower income households in these areas. Dispersed origins and destinations and poor public transportation in these areas make automobile ownership more a necessity than a luxury. Higher average auto emission control costs and more regressive burden patterns are the natural results of these higher levels of car ownership.

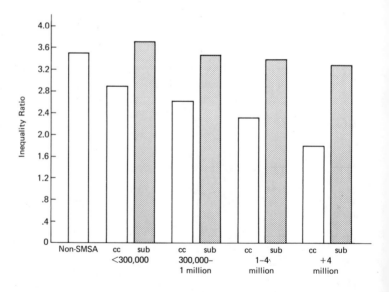

Figure 6-6. Inequality Ratios for Five Population Size Classes in 1990

VARIATION BY REGION

One other division of the 243 SMSAs was analyzed. These urban areas were divided into four regional categories—Northeast, North central, South, and West. Figure 6-7 shows the state boundaries of each of these regions, which correspond to the 1970 census definitions. Appendix B lists the states in each region. The non-SMSA area was not broken down by region in the 1970 census data used in this study so the regional estimates only pertain to urban households.

Average dollar cost burdens for the central cities and suburbs of these four regions are listed in Table 6-9 and shown in a bar graph in Figure 6-8. These figures reveal large regional variations in average cost burdens. Average costs are smallest in the Northeast, followed by the North Central, the South, and the West. The greatest difference is between the Northeast and the other regions, particularly for the central cities. The average annual cost in Northeast central cities is $33.89, compared to $54.80 in the region with the next lowest cost.

The same two factors account for these regional variations in average cost as accounted for variations in costs for individual SMSAs: differences in the income profile of the population, and differences in the cost burden for households in the same income group. The Northeast's low average cost is due both to its greater proportion of lower income households and to its systematically lower costs for households in a given income group.

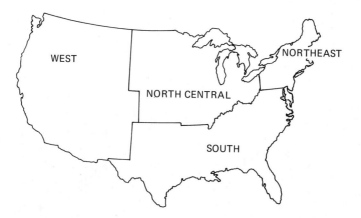

SOURCE: U.S. Bureau of the Census, *Census of Housing: 1970* Metropolitan Housing Characteristics, Final Report HC(2)-1 (Washington, D.C.: U.S. Government Printing Office, 1972), p. xi.

Figure 6-7. Geographic Regions in the United States

Table 6-9.   Average Annual Costs for Four Regions in 1990

Region	*Dollar Costs* *($)*
Northeast	
CC	33.89
SUB	71.42
North Central	
CC	54.80
SUB	75.27
South	
CC	58.43
SUB	72.26
West	
CC	63.43
SUB	76.18

Table 6-10 presents costs by income group for the four regional groups to isolate the regional differences in cost burden for households in the same income group. The differences between the Northeast and the rest of the country are striking, particularly for lower income households living in the central cities. For the lowest income group, the average cost borne by a central city household in the Northeast is barely one-half of the average cost for a

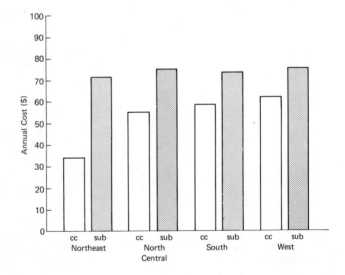

Figure 6-8.   Average Annual Costs for Four Regions in 1990

household in any other region. Northeast households avoid large control cost burdens because they have generally low levels of car ownership. The large, densely populated cities of the Northeast are characterized by close origins and destinations and good public transportation facilities.

Table 6-11 presents information on the regressivity of the cost burden pattern in the four regions. Again the Northeast region stands out. The Northeast has the least regressive cost burden pattern in both the central city and suburban areas. Inequality ratios for each region are presented in Figure 6-9.

Table 6-10. Annual Dollar Costs by Income Group for Four Regions in 1990

Region	<3	3-5	5-7	*Income Group* 7-10	10-15	15-25	25+
Northeast							
CC	9.78	16.95	25.54	36.58	53.46	70.75	94.90
SUB	23.75	37.51	48.37	59.16	78.72	99.44	135.21
North Central							
CC	17.97	31.29	40.94	52.73	72.84	94.10	127.32
SUB	28.89	43.43	53.00	62.14	81.13	101.00	136.38
South							
CC	21.43	37.08	48.43	59.96	81.52	100.87	133.54
SUB	29.44	46.93	57.08	66.90	87.08	104.66	136.26
West							
CC	23.85	38.62	49.30	60.48	81.84	101.85	134.47
SUB	32.84	47.86	57.71	67.50	88.05	106.61	141.16

Table 6-11. Relative Cost Burdens by Income Group for Four Regions in 1990

Region	<3	3-5	5-7	*Income Group* 7-10	10-15	15-25	25+	*Inequality Ratio*
Northeast								
CC	.54	.42	.43	.43	.43	.35	.32	1.7
SUB	1.32	.94	.81	.70	.63	.50	.45	2.9
North Central								
CC	1.00	.78	.68	.62	.58	.47	.42	2.4
SUB	1.60	1.09	.88	.73	.65	.50	.45	3.5
South								
CC	1.19	.93	.81	.71	.65	.50	.46	2.7
SUB	1.63	1.17	.95	.79	.70	.52	.45	3.6
West								
CC	1.32	.97	.82	.71	.65	.51	.45	3.0
SUB	1.82	1.20	.96	.79	.70	.53	.47	3.9

The rankings of the regions in terms of inequality ratios is the same for central city and suburban areas, ranging from the lowest values in the Northeast to the North Central, the South, and the West.

These results indicate that the emission control program will have quite different cost burdens on households in different regions. Costs will be relatively small and fall less heavily on the poor in northeast SMSAs. In contrast, costs will be relatively large and fall more heavily on the poor in western areas. Because the non-SMSA category is not included, these conclusions only refer to urban households. Of course, regional variations in the cost burdens of government programs are common. Whether regional differences of the sort described in this study are relevant for decision making is a political question.

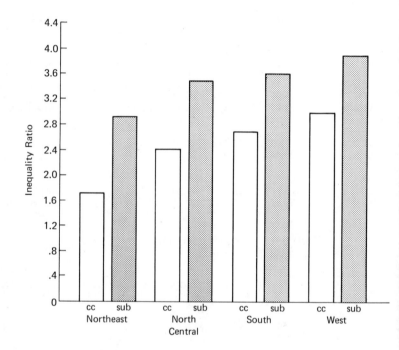

Figure 6-9. Inequality Ratios for Four Regions in 1990

Chapter Seven

Distribution of Air
Quality Benefits

This chapter continues the presentation of empirical results for the current automobile emission control program. In this chapter estimates of the benefits of the plan are presented. The order of presentation is the same as with the costs. National benefits of emission control are first provided and then variations in benefits are given by income group, by urban area, by population size group, and by region of the country.

TOTAL NATIONAL BENEFITS

The concept of national automobile emission control benefits is more ambiguous and more difficult to define than national emission control costs. The measures of benefits used in this study are the total reductions in household exposure to concentrations of the three auto pollutants CO, NO_x and O_x. The measures are calculated by multiplying the concentration benefits in each geographic area by the number of households in the area, and then summing the results for all 487 geographic areas in the study. Average air quality benefits are obtained by dividing these totals by the total number of households in the United States.

Table 7-1 lists total national benefits and average benefits for American households in 1980 and 1990. The last two columns contain the average *percentage* improvement these benefits represent over the 1970 baseline concentration levels. These percentage improvements are represented graphically in Figure 7-1.

Improvements in air quality brought about by the current emission control strategy will be large, particularly in 1990 when the full impact of the program is felt. In 1990, the national benefits of CO concentration reduction will be 673 million household–ppm, or an average of 10.6 ppm per household. This 1990 average CO concentration benefit represents 81 percent of the average 1970 baseline CO concentration. The percentage improvement in CO

Table 7-1.   National Annual Air Quality Benefits of Emission
Controls in 1980 and 1990

Benefits	CO	Air Pollutant O_x	NO_x
Total national benefits (in millions of households–ppm)			
1980	491	449	820
1990	673	540	1175
Average benefits per household			
1980	7.7	7.1	12.9
1990	10.6	8.5	18.5
1970 Average air quality	13.1	12.5	38.3
Percentage improvement from 1970			
1980	59	57	34
1990	81	68	48

Note: Pollutant concentrations are measured as follows

CO: 8 hour annual maximum in parts per million (ppm)

O_x: 1 hour annual maximum in ppm $\times$ 100

NO_x: 24 hour annual average in ppm $\times$ 100

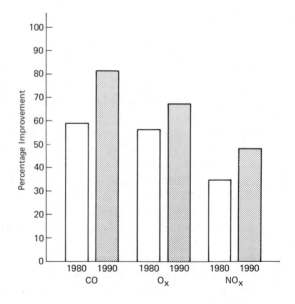

Figure 7-1.   Percentage Improvements in Average Pollutant Concentrations Over 1970 Baseline in 1980 and 1990

concentration is greater than for O_x or NO_x since automobile emissions account for a larger fraction of total CO emissions than of total HC emissions (the precursor of O_x) or total NO_x emissions. But national O_x and NO_x benefits are still substantial. The average 1990 reductions represent a 68 percent improvement for O_x and a 48 percent improvement for NO_x.

Air quality benefits in 1980 are considerably less than the long term 1990 benefits because many of the cars on the road in 1980 will have emission rates similar to uncontrolled (1970) cars. No other 1980 air quality estimates are presented in this chapter because the distributional conclusions from the 1980 figures are virtually the same as those from the 1990 figures; only the general magnitudes are different. Variations in air quality benefits among income groups and geographic areas are discussed in terms of the 1990 long term benefits of the auto emissions control strategy.

VARIATION BY INCOME GROUP

National benefit estimates are of limited usefulness because the auto emission control program will generate vastly different air quality benefits depending upon where a household lives. The rest of this chapter explores these variations in air quality benefits. This section considers whether there are systematic variations in average benefits by income group for the country as a whole.

It is possible to speculate on the variation in average benefits by income group. Since benefits depend upon where a household lives, households in each income will receive the full range of benefits. Some representatives of each income group live in outlying areas where air quality is already very good and thus where benefits are small, while others live in large central cities with very poor air quality and thus large benefits from auto emission control. This spread in benefits applies to poor as well as rich. But since the poor tend to be overrepresented in large central cities, we expect that lower income groups will obtain a *larger* than average share of the physical air quality benefits from auto emission control.

Table 7-2 presents estimates of CO concentration benefits for house-

Table 7-2. National Annual CO Concentration Benefits by Income Group in 1990

Income Group	CO Benefit (ppm)
<3	11.0
3–5	11.0
5–7	11.1
7–10	10.3
10–15	9.9
15–25	10.7
25+	11.4
Pro-poor Ratio	0.96

holds in the seven income groups. For simplicity of exposition, only the CO concentration benefits are used to illustrate variations in air quality benefits; the conclusions would be no different if NO_x and O_x variations were also shown. These average national CO benefits are shown on a bar diagram in Figure 7-2.

Air quality benefits are virtually the same for all income groups. The range in average benefits is only from 9.9 ppm for the $10,000–$15,000 group to 11.4 ppm for the over $25,000 group. The pro-poor ratio (the ratio of benefits for the under $3,000 group to benefits for the over $25,000 group) of 0.96 accurately reflects the similarity of average benefits for households in all income groups.

This similarity of average benefits among income groups was not expected. Since lower income groups are concentrated in the central city areas of most SMSAs, and these central city areas obtain much higher benefits than the suburban areas of the same SMSA, lower income groups were expected to

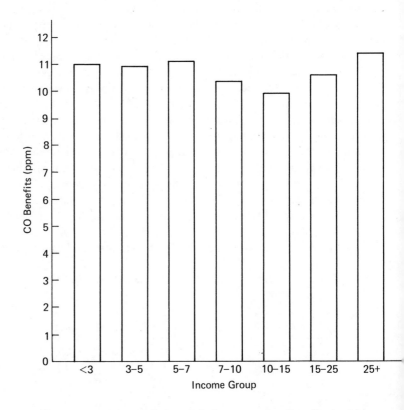

Figure 7-2. National Annual CO Concentration Benefits by Income Group in 1990

experience more benefits than high income groups.[a] However, such an analysis does not take into account the income profile of the non-SMSA area, whose residents experience the lowest air quality benefits of any area in the country. Table 7-3 gives the percentage of American households living in non-SMSA areas for all seven income groups. The income profile of these rural areas is very heavily weighted to low income households. While 30.9 percent of all households live in non-SMSA areas, 40.8 percent of households in the lowest income group live in non-SMSA areas. That percentage declines steadily with income, falling to 16.4 percent for households in the highest income group.

An explanation for the observed constancy in air quality benefits across income groups emerges from these figures on the income profile of non-SMSA areas. Lower income groups are overrepresented in central city areas,

Table 7-3. Percentage of United States Households Living in Non-SMSA Areas by Income Group

Income Group	Percentage Non-SMSA Households
< 3	40.8
3–5	37.8
3–7	35.7
7–10	31.7
10–15	25.6
15–25	19.2
25 +	16.4
Total U.S.	30.9

SOURCE: Bureau of the Census, *Census of Housing: 1970 Metropolitan Housing Characteristics,* Final Report HC(2)-1, (Washington, D.C.: U.S. Government Printing Office, 1972), pp. 1–22, 1–73.

---

[a] The benefit distribution is noticeably more pro-poor for SMSA residents than for all American households.

Income Group	Average CO Benefits for SMSA Households
< 3	17.6
3–5	16.8
5–7	16.5
7–10	14.4
10–15	12.8
15–25	12.8
25 +	13.3
Pro-poor ratio	1.32

where air quality benefits are high, *and* in non-SMSA areas, where air quality benefits are low. These two factors cancel each other out, and an approximately constant distribution of national automobile emission control benefits results.[b]

The national results given in Table 7-2 may, however, understate the air quality benefits of auto emission control to higher income households and overstate the benefits to lower income households. As discussed in Chapter five, the calculations of this study assume that households receive benefits from air quality improvements only in the area in which they live, be it central city or suburb. Some household members will receive benefits from improvements in other urban areas or the other section of their own urban area. In particular, higher income suburban residents are likely to enjoy some of the central city air quality improvements when they enter core areas for work, entertainment, or shopping.

These results also assume that rents do not increase in response to air quality improvements. If renters eventually pay higher rents because of better local air quality, part of the benefits of auto emission control will be transferred from renters to landlords. Since the income profile of landlords is likely to be more skewed to higher income groups than that of renters in most urban areas, housing market adjustments of this sort should decrease the benefits to the lower income groups. The net effect of incorporating household mobility and rental increases in the benefit estimates, therefore, probably would be to make the distribution of national air quality concentration benefits favor the rich.

VARIATION BY URBAN AREA

This section considers variations in benefits for households living in different urban areas and in different population size groups. The first part deals with

---

[b]In the discussion of relative benefits in Chapter Two, another measure of benefits was identified—air pollutant concentration improvement divided by household income. The distribution of benefits among households in different income groups is much more pro-poor when relative benefits are proxied in this way, as the following table shows (the figures are multiplied by 10,000 for readability).

Income Group	CO Concentration Benefits/Household Income
<3	61.2
3–5	27.5
5–7	18.5
7–10	12.1
10–15	7.9
15–25	5.3
25 +	3.8
Pro-poor Ratio	16.1

variations in average benefits in these geographic areas. The second part considers differences in income distributional patterns.

Average Air Quality Benefits

Air quality benefits were calculated in this study for all 487 geographic areas. As was true with cost variations, it is not necessary to present the benefit results for all 487 areas. Table 7–4 lists the 1990 air quality benefits for the same five SMSAs used to illustrate differences in cost burdens in Chapter Six. Bar graphs for CO benefits are shown in Figure 7–3.

Several conclusions emerge from these results. For one thing, the range of benefits of the auto emission control strategy is very large. The CO concentration benefits vary from a 1.5 ppm improvement in the suburbs of Topeka to a 22.7 ppm improvement in the central city of Boston. This result should not, however, be interpreted to mean that Boston residents will have better air quality in 1990 than households in Topeka. In 1990, CO concentrations in Boston are predicted to be 5.4 ppm in the city and 0.6 ppm in the suburbs, while CO concentrations in Topeka are predicted to be 1.1 ppm in the city and 0.4 in the suburbs. Air quality benefits are relatively small in some urban areas because air quality there is already very good, even without emission

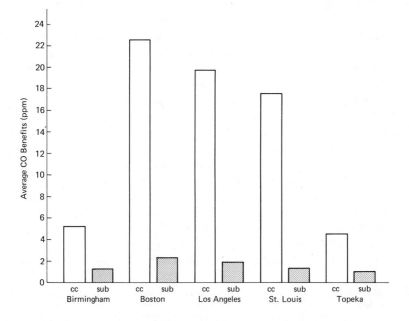

Figure 7–3. Average Annual CO Concentration Benefits in Five SMSAs in 1990

Table 7–4.   Average Annual Air Quality Benefits in Five SMSAs in 1990

SMSA	CO	Central City O_x	NO_x	CO	Suburbs O_x	NO_x
Birmingham	5.4	3.6	18.4	1.6	1.0	11.3
Boston	22.7	17.2	33.9	2.3	1.8	16.1
Los Angeles	19.9	14.7	31.8	2.0	1.4	13.8
St. Louis	17.9	13.2	30.4	1.7	1.1	11.9
Topeka	4.5	3.0	17.2	1.5	0.9	11.0

controls. Cities like Boston and Los Angeles, on the other hand, obtain significant air quality improvement benefits from federal automobile emission controls because uncontrolled air quality is so poor.

A second conclusion emerging from these results is that air quality benefits are generally small in suburban areas, even suburban areas of large urban centers like Boston and Los Angeles. The suburban CO concentration benefits in Table 7-4 only range from 1.5 ppm for Topeka to 2.3 ppm for Boston. Benefits are small because suburban areas are generally characterized by low emissions densities and therefore low uncontrolled pollutant concentrations.[c] The auto emission control program cannot markedly improve on already good suburban air quality.

The figures in Table 7-4 also suggest a relationship between air quality benefits and SMSA population. Benefits should be larger in larger *cities,* but roughly the same for all suburban areas. In Table 7-5 the average air quality benefits are listed for the same five population size categories used to analyze cost differences—four SMSA size classes and the one nonurban group. Bar graphs of average CO benefits for these five size classes are given in Figure 7-4.

The figures in Table 7-5 confirm the expectations about the relation between SMSA size and central city air quality benefits. Air quality benefits increase as the SMSA size increases. The differences in central city air quality benefits among the four SMSA size classes are dramatic. The average 1990 CO benefit for central city households ranges from 69.9 ppm in SMSAs with more than four million persons to 17.2 ppm for SMSAs with one to four million persons, 8.0 ppm for SMSAs with 300,000 to one million persons, and 5.7 ppm for SMSAs with fewer than 300,000 persons.

These results indicate that benefits of auto emission control will be much greater for households in the central cities of the few largest SMSAs than

---

[c]This study may understate the suburban improvements and overstate the central city improvements for O_x. Virtually all modeling procedures, including the one used in this study, are poorly suited to estimate O_x concentrations. Oxidants are produced by complicated atmosphere reactions, which take place over several hours. Because of this time lag, the highest concentrations of oxidants may exist not in central city areas where emissions density is greatest but in suburban areas in the path of the prevailing winds.

Table 7-5. Average Annual Air Quality Benefits for Five Population Size Classes in 1990

Population Size	Central City CO	Central City O_x	NO_x	Suburb CO	Suburb O_x	NO_x
4 million +	69.6	59.5	52.6	2.0	1.4	14.2
1-4 million	17.2	13.0	27.6	1.9	1.3	13.0
300,000-1 million	8.0	5.6	20.3	1.7	1.2	12.3
<300,000	5.7	3.9	18.3	1.6	1.0	11.3
Non-SMSA				1.5	0.9	11.0

for households in other central city areas. It is somewhat misleading, however, to refer to these large cities as a small group since households in these five central cities account for about 10 percent of the total United States population. The air quality benefits for the four suburban areas and the non-SMSA area are very similar to one another, although the same general trend toward increasing benefits with increasing SMSA size is evident for the suburban figures.

Income Distributional Patterns

The distribution of benefits in American *urban areas* is pro-poor because lower income groups are overrepresented in central city areas where air quality benefits are high. National benefits were found to be roughly the same for all income groups because the poor are also overrepresented in nonurban

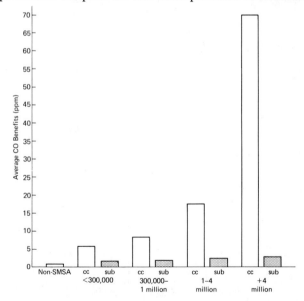

Figure 7-4. Average Annual CO Concentration Benefits for Five Population Size Classes in 1990

areas. But the question remains whether there are systematic differences in the distributional patterns *among* SMSAs. This section considers variations in income distributional profiles in the sample of five SMSAs and in the SMSA size classes.

Table 7-6 lists the annual CO concentration benefits by income group for the five SMSAs. The pro-poor ratios for each SMSA are listed in the last column and graphed in Figure 7-5. As expected, the benefits of the federal auto emission control scheme are distributed in a pro-poor manner within all SMSAs. But there is considerable variation, as can be seen by comparing pro-poor ratios. In these five SMSAs, the pro-poor ratio varies from 1.0 for Topeka to 2.5 for St. Louis. Thus, in Topeka the highest and lowest income groups will enjoy approximately the same air quality benefits, while in St. Louis members of the lowest income group will on average receive 2.5 times the CO concentration benefits received by members of the highest income group. This variation is the result of different patterns of concentration of lower income groups. In Topeka, households in different income groups are distributed quite evenly between central city and suburban areas, while in St. Louis lower income groups are heavily concentrated in the city of St. Louis.

Table 7-6. Annual CO Concentration Benefits by Income Group
for Five SMSAs in 1990

| | Income Group | | | | | | | |
SMSA	<3	3-5	5-7	7-10	10-15	15-25	25+	Pro-poor Ratio
Birmingham	3.5	3.5	3.3	3.2	3.1	2.9	2.4	1.5
Boston	10.2	10.0	9.3	7.9	6.3	5.3	4.5	2.3
Los Angeles	12.3	11.9	11.2	10.2	9.5	9.6	10.7	1.1
St. Louis	9.3	8.9	8.0	6.4	4.9	4.2	3.7	2.5
Topeka	4.2	4.2	4.1	4.0	4.0	4.0	4.1	1.0

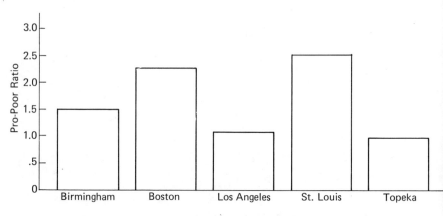

Figure 7-5. Pro-poor Ratios in Five SMSAs in 1990

The results for these five SMSAs suggest systematic differences in income distributional patterns by population size class. Larger SMSAs should have a more pro-poor benefit distribution because lower income households are more concentrated in central city residential areas and because, as shown in the last section, the gap between central city and suburban benefits is very great for large SMSAs. In smaller SMSA the distribution of benefits will be less pro-poor, both because the difference in air quality benefits between central city and suburban areas is relatively small and because lower income households are less concentrated in central city residential areas.

Table 7-7 presents evidence to test for systematic difference in distributional patterns among SMSA size classes. The pro-poor ratios for the five size classes are illustrated by bar graphs in Figure 7-6. These results show the expected trend in income distributional patterns. Benefits become more pro-poor as population size increases. Thus, the air quality benefits from the auto emissions control program will be both larger and distributed in a more pro-poor manner in the larger SMSAs.

The benefit trends by SMSA size class seem to reinforce the cost trends analyzed in Chapter Six. Costs were found to be lower and to fall less heavily on the poor in large SMSAs. With benefits greater and accruing more than proportionately to the poor at the same time, these large urban areas have the best of both worlds. Households there get large benefits while paying relatively small costs. Moreover, lower income groups in these areas are better off compared to their counterparts in smaller SMSAs and non-SMSA areas because the costs are less regressively distributed and the benefits are markedly pro-poor.

Table 7-7. Annual CO Concentration Benefits by Income Group for Five Population Size Classes in 1990

SMSA Size	<3	3-5	5-7	7-10	10-15	15-25	25+	Pro-Poor Ratio
4 million +	49.8	48.7	47.7	41.6	34.4	31.1	30.3	1.6
1-4 million	10.4	9.9	9.3	8.1	6.8	6.1	5.9	1.8
300,000-1 million	5.3	5.1	.50	4.7	4.4	4.2	4.0	1.3
<300,000	4.1	4.0	3.9	3.8	3.8	3.9	3.9	1.1
Non-SMSA	1.5	1.5	1.5	1.5	1.5	1.5	1.5	1.0

The columns <3 through 25+ fall under the heading *Income Group*.

Figure 7-6. Pro-poor Ratios in Five Population Size Classes in 1990

VARIATION BY REGION

Although the auto emissions control strategy is currently a national program, regional differences in air quality benefits are still of interest to policy makers. Table 7-8 presents estimates of the average 1990 SMSA air quality benefits for households in four regions of the country—Northeast, North Central, South, and West. The CO concentration benefits are shown in Figure 7-7. The non-SMSA category was not broken down by region, so these regional estimates are only for SMSA households. If the proportion of non-SMSA households differs by region, these SMSA figures may not capture some of the true inter-regional differences.

There is considerable regional variation in the benefits of the control strategy. The regions listed in order of increasing central city benefits are Northeast, North Central, West, and South. Little regional variation in benefits exists for suburban areas. The large benefits for Northeastern households reflect the greater proportion of older, densely settled cities in the Northeast, which gain a great deal from reductions in auto emissions. In contrast, the newer, less dense cities of the South have good air quality without stringent auto emissions controls.

Table 7-9 presents regional averages of the income distributional impacts of emission control benefits. Average pro-poor ratios are shown in Figure 7-8. Benefits are distributed in the most pro-poor manner in the North Central region, followed closely by the Northeast region. The distributional patterns in the South and West are approximately the same and substantially less pro-poor than the patterns in the other two regions.

Table 7-8. Average Annual Air Quality Benefits for Four Regions in 1990

| Region | Central City | | | Suburb | | |
	CO	O_x	NO_x	CO	O_x	NO_x
Northeast	64.2	55.2	48.3	2.1	1.6	14.4
North Central	21.1	16.4	29.6	1.8	1.2	12.5
South	7.2	5.1	19.4	1.7	1.1	11.8
West	15.4	11.5	26.5	1.8	1.2	12.5

Table 7-9. Annual CO Concentration Benefits by Income Group for Four Regions in 1990

| Region | Income Group | | | | | | | Pro-poor |
	<3	3-5	5-7	7-10	10-15	15-25	25+	Ratio
Northeast	40.0	38.7	37.0	31.1	26.8	26.4	26.6	1.5
North Central	14.5	13.7	13.4	11.4	9.8	9.4	8.4	1.7
South	5.0	4.9	4.7	4.4	4.1	4.0	4.2	1.2
West	9.4	8.8	8.6	7.8	7.0	7.0	7.8	1.2

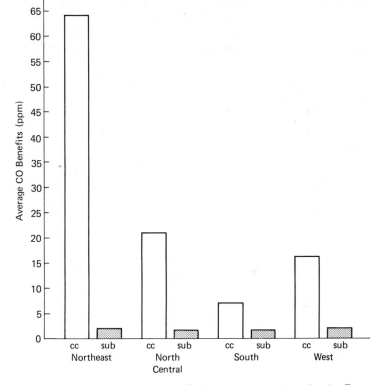

Figure 7-7. Average Annual CO Concentration Benefits in Four Regions in 1990

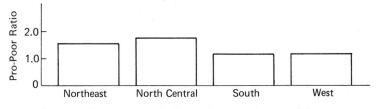

Figure 7-8. Pro-poor Ratios in Four Regions in 1990

Chapter Eight

Alternative Automobile Emission Control Strategies

The bulk of this study deals with the impacts of the current automobile emission control program mandated by Congress in the 1970 Clean Air Act Amendments. The last two chapters provided detailed estimates of the cost and benefit impacts the current scheme will have on households in various income and geographic groups. Some of these consequences are disturbing, particularly on the cost side. Control costs were found to be a larger fraction of income for households in lower income groups. Households living in suburban areas, small urban areas, and nonurban areas—which make up two-thirds of United States households—do poorly under the current scheme. Households in these areas gain quite modest air quality benefits while paying large costs. Lower income groups in these areas fare particularly poorly since the costs fall quite heavily upon them under the current scheme.

The analysis of alternative automobile emission control strategies in this chapter is a natural extension of the detailed analysis of the current control program. The objective in this chapter is not to evaluate all possible options but rather to compare the economic impacts of plausible alternatives with an eye toward identifying strategies which avoid the adverse distributional impacts of the current scheme. A more complete analysis of alternatives would include comparisons with nonautomobile emission control strategies, such as reductions in emissions from power plants and other stationary sources, as well as with strategies that decrease concentrations by reducing automobile travel rather than auto tailpipe emissions.

Many modifications in the current control plan can be envisioned. One option is to relax the final emission standards on the theory that the final standards either cannot be met in a technological sense or can only be met at greatly increased costs. Alternatively, some commentators have advised keeping the present final standards but postponing them for a number of years. The theory behind the postponement option is that the current timetable will result in quite large control costs in the short run. But with postponement, it is argued, the costs of meeting the final standards can be reduced substantially as better technology is used and auto manufacturers realize the cost advantages of production experience with the new technologies.

Suggestions to make the automobile emission control program more stringent have been less frequent. Virtually no commentators have recommended decreasing the final emission standards below the current levels. But some have suggested adding a requirement for an inspection-maintenance (I-M) system to the current emission standard timetable. States would inspect cars for compliance with emission control standards and would require high polluters to bring their cars down to the standards. This I-M program is designed to combat the deterioration of emission controls as cars are driven. Several states are now planning to adopt an I-M program to help meet federal ambient air quality standards in their major urban areas. Thus far the federal government has not required an I-M program.

The most promising suggestion in light of the distributional analysis presented in this study is to allow the emission standards to vary for cars driven or owned in different geographic areas. Many urban areas and virtually all non-urban areas can obtain good air quality in 1990 with relatively lenient automobile emission controls. Recognition of this fact has led to suggestions for a multicar scheme, which would allow the severity of emission control to vary with the severity of the urban area's pollution problem.

It is clearly impractical to have a great many auto emission standards, since automobile production as well as administration and enforcement of emission standards under multicar schemes are more costly and problematic. The most common proposal is for a two car scheme in which small urban and non-urban areas are subject to a single set of more lenient standards while the current emission standards remain in force in the relatively few urban areas with substantial pollution problems.

This chapter compares the current scheme with four alternatives selected to bracket the possible options from more lenient to more stringent auto emission controls. These alternatives are the lenient scheme, the stringent scheme, and two multicar schemes—the two car scheme, and the three car scheme.

The lenient strategy would alter the national new car emission standards so that the 1973 emission standards apply to all cars manufactured in the future. These 1973 new car standards represent roughly a 25 percent reduction in emissions from the 1970 level for all three pollutants. No postponement strategy is considered in this chapter, primarily because the concern in this chapter is with the long term impacts of the control strategies. All costs and benefits are calculated for 1990. A postponement strategy primarily affects the short run and medium run impacts of control. Of course a more complete analysis of alternatives would include postponement strategies.

The second alternative strategy, the stringent strategy, retains the current new car emission standards, but also imposes a national I-M program. All cars would be inspected for high emissions and owners of high emitters would be forced to bring their cars in line with the standards. This I-M program would be designed primarily to reduce emissions from older cars whose emission control systems may have deteriorated.

Both multicar schemes are combinations of the three national schemes (the two options and the current strategy). The two car scheme applies the lenient emission control standards to all cars driven in relatively unpolluted areas and the current standards to cars driven in polluted areas. The three car scheme identifies a group of the most highly polluted areas and adds an additional requirement for inspection-maintenance program in those areas.

The specific characteristics of these alternative schemes are given in the next section. The following section compares the national cost-effectiveness of alternatives. While this study emphasizes the distributional implications of automobile emission control, the impact of various alternatives on economic efficiency is also important in judging their overall merits. The final section compares alternatives from the standpoint of their impacts on households in different income groups.

CHARACTERISTICS OF ALTERNATIVE STRATEGIES

This section presents the cost and emissions characteristics of the car fleet in 1990 under the five alternative schemes. For the three national schemes, these details include the cost and emission characteristics for cars of various age groups in 1990 if each strategy were in force. The two multicar schemes are characterized by identifying the geographic areas subject to various requirements.

National Strategies

The cost and emissions characteristics of cars under the lenient, current, and stringent schemes are given in Table 8-1. Costs are broken down into three categories—ownership costs, fuel costs, and repair and maintenance costs—and are presented for the six car age groups used in other cost analyses. The emissions (in grams per mile) of CO, HC, and NO_x for these car age groups are presented in the right-hand side of the table.

The information in Table 8-1 for the current strategy is the same as that used for the detailed analyses of costs and benefits given in Chapters Six and Seven. The figures for the other two schemes are best viewed as reasonable guesses of the characteristics of the 1990 car fleet under the lenient and stringent alternatives.

Costs for the lenient scheme were derived from the National Academy of Sciences estimates for meeting the 1973 standards.[1] These estimates

Table 8-1.   Costs and Emissions Characteristics of Cars Under
National Automobile Emission Control Strategies

Car Age	Owner Cost ($/yr)	Costs Fuel Cost (¢/mi)	Maintenance Cost ($/yr)	Emissions CO (g/mi)	HC (g/mi)	NO_x (g/mi)
			Lenient Strategy			
New	14.44	.106	5	23.4	3.39	2.19
1	11.25	.106	5	27.6	3.54	2.42
2	8.81	.106	5	30.9	3.70	2.58
3-4	6.05	.106	5	32.5	3.83	2.62
5-7	3.30	.106	5	34.4	4.02	2.69
8+	1.51	.106	5	36.0	4.21	2.75
			Current Strategy			
New	50.41	.175	10	2.21	.47	.30
1	39.59	.175	10	2.67	.51	.35
2	30.99	.175	10	2.98	.55	.40
3-4	21.31	.175	10	3.45	.61	.47
5-7	11.64	.175	10	4.16	.68	.57
8+	5.30	.175	10	4.92	.76	.68
			Stringent Strategy			
New	50.41	.175	10	2.21	.47	.30
1	39.59	.175	11	2.44	.50	.32
2	30.99	.175	12	2.66	.53	.35
3-4	21.31	.175	15	3.10	.58	.41
5-7	11.64	.175	20	3.10	.58	.41
8+	5.30	.175	20	3.10	.58	.41

Note: Pollutant emissions are listed for an average speed of 15 miles per hour.

may overstate the costs to meet the same standards in the 1980s because of technological developments and learning curve phenomena which decrease the costs of meeting given standards over time. The Academy figures were used without alteration because no quantitative estimates of these effects are available. The emission rates by car age listed in Table 8-1 for the lenient scheme are based on EPA estimates of 1973 new car emission rates and deterioration factors relating emission rates to car age.[2]

The costs and emissions estimates for the stringent scheme involve even more guesswork that those for the lenient strategy. The inspection-maintenance program required under the stringent scheme is assumed to increase annual maintenance costs by amounts ranging from $1 per year for one year old cars to $10 per year for cars eight years and older. These increased maintenance costs are assumed to generate improvements in per mile emissions ranging from around 5 percent of the current scheme levels for one year old cars to around 30 percent for cars in the oldest category. These costs and emissions improvement figures seem plausible, although the actual costs and emissions changes would depend on the severity of the inspection and maintenance requirements and the effectiveness of enforcement provisions.

Note that the cost estimates for the stringent strategy disregard the nonmaintenance costs associated with the I-M strategy, which include investment costs for the testing facilities and equipment and labor costs for testing, administration, and enforcement. Neglecting these costs will bias the cost-effectiveness and income-distributional analyses in favor of the stringent scheme. In fact, the stringent scheme will probably be less cost-effective and, because these other costs would probably be financed by a per car inspection charge, fall more heavily on the poor than estimated in this chapter.

Multicar Strategies

As mentioned, both multicar strategies considered in this study are combinations of the three national schemes. Under the two car strategy, lenient emission standards apply to cars owned in relatively unpolluted areas while the current standards apply to cars owned in polluted areas. The three car strategy further subdivides polluted areas into those with moderate pollution problems, in which the current standards apply, and those with severe pollution problems, in which the stringent requirement for an inspection-maintenance system is added.

The major problem in characterizing both the two car scheme and the three car scheme is to divide the country into unpolluted, polluted, and severely polluted areas. Since the uncontrolled air pollutant concentrations of the 487 geographic areas in the study form a continuum, from very low concentrations to extremely high concentrations, the choice of dividing lines between unpolluted, polluted, and severely polluted is clearly somewhat arbitrary. The benchmarks used to place these divided lines are the federal ambient air quality standards for the three auto pollutants.[a]

Under the two car strategy, current standards are assumed to be in force in the 43 SMSAs whose central cities exceed the ambient air quality standards for any of the auto pollutants in 1990 with the *lenient* emission control strategy in force. Table 8-2 shows that these 43 large, dense SMSAs account for 36.6 percent of the households in the United States. Cars of the remaining 63.4 percent of American households, living in the other 200 SMSAs and in non-SMSA areas, will be subject to the lenient standards.

The three car scheme divides the 43 SMSAs into a group of 19 that are severely polluted and a group of 24 that are polluted. The 19 SMSAs are

---

[a]The federal primary ambient air quality standards, measured in the same units as in this study, are the following:[3]

CO : 9– 8 hour annual maximum concentration in ppm
NO_x : 50–24 hour annual average concentration in 1000 $\times$ ppm
O_x : 8– 1 hour annual maximum concentration in 100 $\times$ ppm

Table 8-2.   Division of American Households Under Multi Car
Automobile Emission Control Strategies

		Households	
Applicable Standard	*Geographic Areas*	*Number (millions)*	*Percentage of U.S.*
Two Car Strategy			
Current standards	43 SMSAs	23.25	36.6
Lenient standards	200 SMSAs	20.61	32.5
Lenient standards	non-SMSA	19.59	30.9
Total U.S.		63.45	100.0
Three Car Strategy			
Stringent standards	19 SMSAs	15.33	24.2
Current standards	24 SMSAs	7.92	12.4
Lenient standards	200 SMSAs	20.61	32.5
Lenient standards	non-SMSA	19.59	30.9
Total U.S.		63.45	100.0

those whose central cities exceed the national ambient standards in 1990 even
with the *current* emission standards. These 19 SMSAs have 15.3 million households, which is almost one-quarter of the United States total. In general, these
SMSAs are the largest and most densely populated areas. The 24 SMSAs in the
second group meet the ambient standards in 1990 with the current auto
controls. The third group in the three car strategy is the same as the second
group in the two car strategy, consisting of the other 200 SMSAs and the non-
SMSA areas.

COST-EFFECTIVENESS COMPARISONS

This section evaluates the five alternative strategies from the standpoint of their
impact on national income. As mentioned several times in previous chapters,
since benefits are not measured in dollars in this study, it was not possible to
determine whether the benefits of the current scheme outweigh the costs in an
economic efficiency standpoint. Similarly, the net benefits of alternatives cannot
be calculated in dollar terms. But the costs and the air quality benefits of alternatives can be compared to provide judgments about the cost-effectiveness of
the various strategies.

Information on the 1990 national costs and benefits of alternative
schemes are given in Table 8-3 and Table 8-4, respectively. These national
totals are the sum of costs or benefits for households in all 243 SMSAs and the
non-SMSA group. Total control costs are broken down by the three major cost
components. Benefits are measured by reduced exposure to the three automobile pollutants. Figures 8-1 and 8-2 summarize the national cost and benefit
impacts for the five schemes. Figure 8-1 shows the average costs per household,

while Figure 8-2 shows the average CO benefits per household as a percentage of the 1970 baseline level.

These national cost and benefit figures provide the data to compare the cost effectiveness of the five strategies. The measure of cost effectiveness used in these comparisons is average CO concentration improvement per dollar average cost, multiplied by 1,000 for readability. For example, since the lenient strategy obtains a 5.32 ppm per household improvement in CO concentration over the baseline level at an average cost of $28.23 per household, the cost-

Table 8-3. National Annual Costs of Alternative Strategies by Cost Component in 1990

Strategy	Ownership	Operating	Stockholder and Taxpayer	Total	Cost Per Household
		Control Costs ($1,000s)			
Lenient	416	1,292	121	1,829	$28.83
Two car	791	1,644	147	2,582	$40.73
Three car	791	1,748	147	2,686	$42.37
Current	1,463	2,316	180	3,959	$62.40
Stringent	1,463	2,841	180	4,484	$70.67

Table 8-4. National Annual Air Quality Benefits of Alternative Strategies in 1990

Strategy	National Benefits (millions of) household-ppm)	Per Household Benefits (ppm)	Benefits as Percentage of 1970 (%)
	CO		
Lenient	338	5.3	41
Two car	622	9.8	75
Three car	627	9.9	75
Current	673	10.6	81
Stringent	681	10.7	82
	O_x		
Lenient	304	4.8	38
Two car	513	8.1	65
Three car	517	8.1	65
Current	540	8.5	68
Stringent	555	8.6	69
	NO_x		
Lenient	607	9.6	25
Two car	923	14.6	38
Three car	933	14.7	38
Current	1175	18.5	48
Stringent	1203	19.0	49

effectiveness of the lenient strategy is 185 ($\dfrac{5.32}{28.23}$ • 1000). For simplicity, only CO concentration is used. The same conclusions would result if effectiveness were measured by concentrations of either of the other two auto pollutants or by some average of the three air quality benefits.

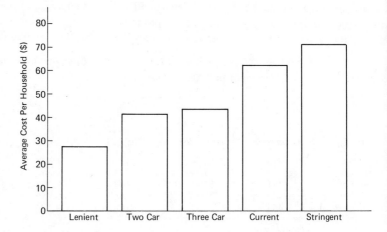

Figure 8-1. National Average Annual Cost Per Household for Alternative Strategies in 1990

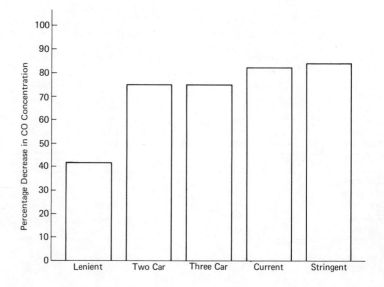

Figure 8-2. National Average Percentage Decrease in CO Concentration over the 1970 Baseline for Alternative Strategies in 1990

One way to compare the other schemes to the lenient scheme's cost-effectiveness would be to contrast the average cost-effectiveness of each strategy. But averages would be misleading. The auto emission control strategies are *alternatives* for one another and represent a series of increasingly stringent control plans. The incremental nature of the schemes can be seen by restating the nature and geographic coverage of controls under each scheme.

Lenient strategy: 1973 emission standards apply to all cars sold in the United States

Two car strategy: Same as lenient strategy except that 1977 emission standards apply to cars sold in 43 polluted SMSAs

Three car strategy: Same as two car strategy except that cars driven in the 19 most polluted SMSAs are also subject to an I-M program

Current strategy: 1977 emission standards apply to all cars sold in the United States

Stringent strategy: Same as current strategy with all cars also subject to the I-M program

As alternatives, the proper comparison is not average cost-effectiveness but *incremental* cost-effectiveness—i.e., the additional benefits (measured by air quality improvements rather than dollars) divided by the incremental costs.

Table 8-5 lists the incremental benefits, incremental costs, and the quotient, incremental cost-effectiveness for the five control schemes. This table reveals great differences in the cost-effectiveness of the various strategies. Incremental cost-effectiveness ranges from 16 for the stringent strategy to 376 for the two car strategy. These figures are not precise, but they do indicate the very different orders of magnitude of the incremental costs of implementing each strategy in lieu of a strategy less stringent. According to these estimates, the lenient scheme and the two car scheme are quite cost-effective, while the three more stringent control measures all involve decreasing benefits in comparison to costs.

Table 8-5. Incremental Cost-Effectiveness of Alternative Strategies in 1990

Strategy	Incremental CO Benefit (ppm)	Incremental Cost ($)	Incremental Cost-Effectiveness (ppm/$ X 1000)
Baseline	0	0	0
Lenient	5.32	28.83	185
Two car	4.48	11.90˙	376
Three car	.07	1.64	43
Current	.74	20.03	37
Stringent	.13	8.27	16

The decreasing cost-effectiveness beginning with the three car strategy only relates to a relatively small range of benefits. Figure 8–3 graphs the cost-effectiveness of each scheme against the range of benefits it encompasses. This graph illustrates that cost-effectiveness is relatively high over most of the range of per household benefits, because the two car strategy achieves over 93 percent of the benefit achievable with any of the five strategies. The area of very low cost-effectiveness, representing the three car strategy, the current strategy, and the stringent strategy, is confined to the last 7 percent of the CO benefits.

The superior cost-effectiveness of the two car strategy was expected. The detailed results presented in Chapter Six and Seven for the current strategy revealed large geographic variations in costs and benefits. Large central cities obtained large benefits from stringent auto emission controls while paying relatively low average costs. Smaller urban areas and rural areas achieved very modest benefits while paying large control costs. By mandating stringent auto emission controls in large, heavily polluted urban areas and lenient controls in smaller, less polluted urban areas and rural areas, the two car strategy concentrates stringent controls in areas where benefits are high and costs are low. Households in smaller urban and rural areas under the two car strategy still receive small air quality benefits, but at greatly reduced control costs.

The cost-effectiveness of the two car strategy would be even greater if the geographic differentiation were more specific than assumed in this study. For example, the stringent automobile emission controls could be mandated only for cars driven in the central city areas of impacted SMSAs. In that case,

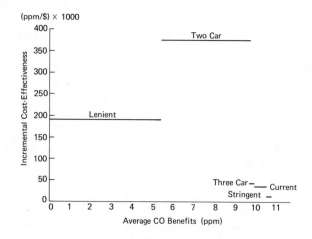

Figure 8–3. Relation Between Incremental Cost-Effectiveness and Average CO Benefits as Increasingly Stringent Strategies are Employed

suburban residents in the large, polluted urban areas would be spared the large costs of stringent controls which only generate modest air quality benefits. But the need for mobility between subareas of a single urban area and the difficulties of enforcing such specific controls are likely to override the advantages of such selective standards. Indeed, some have suggested a less specific two car scheme in which all cars driven in a *state* with a polluted urban area would be subject to the more stringent standards. Legislating standards by state may ease the administrative burden on a multicar scheme but would increase the control costs of the program and reduce its cost-effectiveness since many of the cars in states with polluted SMSAs are driven in areas without pollution problems. Since the problems of administering a multicar scheme based on SMSAs are likely to be minor, there seems no compelling reason to make the standards statewide.

The two car scheme also will affect the regional variation in average costs and benefits. The analysis in Chapters Six and Seven indicated that northeast urban areas fare relatively well under the current scheme, receiving large benefits and bearing small costs. This pattern will tend to continue under the two car strategy since most of the heavily populated northeast urban areas will be subject to the stringent emission controls. But the less desirable situation for many cities in the South and West will be improved under the two car strategy. Many of the newer, smaller, more spread out cities in these regions will be subject to the more lenient auto emission controls, which will greatly reduce emission control costs while only sacrificing modest air quality benefits. The large regional disparities observed for the current scheme therefore will tend to be diminished under the two car strategy.

The two strategies using the I-M program, the three car strategy and the stringent strategy, fare poorly by the yardstick of economic efficiency. Both strategies generate small incremental benefits at relatively large incremental costs, with the quotient, cost-effectiveness, therefore quite low. Since the costs to run the testing facilities and administer and enforce the I-M program have been omitted from the cost side of this analysis, a more complete evaluation of the I-M program probably would generate an even more pessimistic judgment.

A word should be said about the lenient scheme. The sizable benefits under the lenient scheme are achieved at rather low cost. Since the cost of achieving the lenient (1973) standards will probably be less than assumed in these calculations, perhaps dramatically less, the cost-effectiveness of the lenient strategy is probably even greater than estimated here. The results for the two car strategy suggest that the incremental benefits of stringent controls in polluted urban areas are worth the extra cost. But to answer this economic efficiency question in a more definitive way, it would be necessary to know the dollar value of these additional benefits.

INCOME-DISTRIBUTIONAL COMPARISONS

This section compares the strategies from the standpoint of their equity impacts—

their relative impacts on households in different income groups. The first part discusses the cost impacts and the second part discusses the benefit impacts.

Cost Impacts

It is possible to make a priori judgments about the national distribution of costs under the alternative automobile emission control strategies. As discussed in Chapter Six, the current scheme has a regressive cost burden pattern —costs are a larger fraction of income for lower income groups. Neither of the national alternatives should reduce this regressivity. The cost burden pattern for the lenient strategy should be similar to the current scheme, although of course the general level of control costs for each income group is much lower. The stringent scheme should have the most regressive cost distribution, as well as the highest average burden, since that strategy increases maintenance costs on older cars. These added costs make the cost burden more regressive because older cars are more likely to be owned by lower income households. The increase in regressivity over the current scheme should not be too large because the total increase in costs of the national I-M program is relatively small (average costs only increase from $62.40 to $70.67).

The two car strategy, in contrast, promises to yield improvements in the income distributional impacts. Reducing auto ownership and operation costs for households in medium sized and small SMSAs and the non-SMSA areas will lessen burdens on low income groups more than high income groups. There are several reasons for expecting this pattern. For one thing, the large, heavily polluted SMSAs have larger percentages of high income families than the country as a whole, because of the generally better economic climate in these SMSAs. Table 8-6 shows, for each income group, the percentage of households living in the 43 polluted SMSAs subject to the stringent emission controls under the two car strategy. The percentage of households living in these 43 urban areas rises

Table 8-6. Percentage of Households in 43 Polluted SMSAs for Income Groups in the United States

Population Group	Income Group						
	<3	3-5	5-7	7-10	10-15	15-25	25+
43 Polluted SMSAs	29.0	30.3	31.9	34.7	40.4	48.5	52.0
Remainder of Country	71.0	69.7	68.1	65.3	59.6	51.5	48.0
Total	100.0	100.0	100.0	100.0	100.0	100.0	100.0
Number of Households (millions)	11.5	7.2	7.6	12.0	14.3	8.4	2.5

steadily with income. While less than 30 percent of households in the lowest income group live in these polluted SMSAs well over 50 percent of the highest income group reside there.

Another reason for expecting the two car strategy to be less regressive than the current strategy is that mandating reductions in control costs for smaller urban and nonurban areas will relieve low income households of substantial cost burdens. The costs of patterns for the current scheme, reported in Chapter Six, were found to fall quite heavily on the poor in these small urban and nonurban areas because low income households living there have quite high levels of car ownership. Low income households in the very large central cities, however, are much more likely to avoid auto emission control costs by having low auto ownership rates. For example, the average cars per household for Boston city households earning less than $3,000 per year is 0.25 compared to 0.61 for Topeka households in the same income group.

The three car strategy should have a somewhat more regressive cost pattern than the two car strategy for the same reason that the stringent strategy will be more regressive than the current scheme: the increased maintenance costs for I-M fall more heavily on lower income households. But the adverse distributional impacts will be muted in the case of the three car scheme because the I-M plan only applies to 19 large SMSAs where, as noted above, low income households are likely to have relatively low car ownership rates.

Table 8-7 presents the income distributional results. The top half shows the annual dollar costs for households in different income group under

Table 8-7. National Annual Costs and Relative Cost Burdens by Income Group of Alternative Strategies in 1990
Dollar Costs ($)

Strategy	*<3*	*3-5*	*5-7*	Income Group *7-10*	*10-15*	*15-25*	*25+*
Lenient	11.29	18.51	23.12	27.86	36.51	45.17	62.70
Two car	14.07	23.27	29.89	37.50	52.43	69.82	97.73
Three car	14.69	24.08	30.94	39.04	54.54	72.90	105.38
Current	23.98	39.93	50.25	60.39	79.79	98.38	130.63
Stringent	28.94	46.55	57.93	69.47	89.43	109.09	140.71

Relative Costs (% of income)

Strategy	*<3*	*3-5*	*5-7*	Income Group *7-10*	*10-15*	*15-25*	*25+*	*Inequality Ratio*
Lenient	.63	.46	.39	.33	.29	.23	.21	3.0
Two car	.78	.58	.50	.44	.42	.35	.33	2.4
Three car	.82	.60	.52	.46	.44	.36	.35	2.4
Current	1.33	1.00	.83	.71	.64	.49	.44	3.1
Stringent	1.61	1.16	.97	.82	.72	.55	.47	3.4

each of the five strategies. The bottom half shows the relative costs (costs as a percentage of income).

All emission control schemes are regressive, since the percentage of income taken by control costs always decreases with income. But there are marked differences in the level of regressivity. Inequality ratios (the ratios of the relative burden for the lowest income group to the relative burden for the highest income group) are listed in the last column of Table 8-7 and shown on a bar graph in Figure 8-4. As expected, the stringent strategy is the most regressive scheme, with an inequality ratio of 3.4. The lenient scheme and the current scheme have similar regressive cost burdens: in both strategies the relative burden of households in the lowest income group is about three times the relative burden of households in the highest income group.

The most striking result of Table 8-7 is the difference in the regressivity of the multicar schemes on the one hand, and the national schemes on the other. For both multicar schemes, the inequality ratio is only 2.4. Relaxing standards in relatively small urban areas and nonurban areas clearly reduces the burden of automobile emission controls on low income households. The similarity of the inequality ratio for the two car and three car schemes suggests that adding an I-M program to the most polluted areas does not affect the cost burden pattern.

This cost analysis indicates that the automobile emission control program can be made both less costly and less burdensome on the poor by adopting a multicar strategy. However, the costs of these multicar strategies are still quite large and their impact on the income distribution is still not desirable. Under both multicar strategies the annual costs are around $2.6 billion in 1990 and the relative burden of households in the lowest income group is on average over

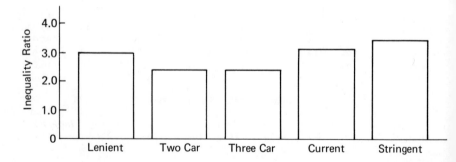

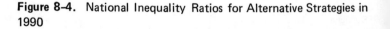

Figure 8-4. National Inequality Ratios for Alternative Strategies in 1990

twice the relative burden of households in the highest income group.

Benefit Impacts

In comparing the cost-effectiveness of alternative schemes, it was found that the four most stringent strategies—the two car scheme, the three car scheme, the current scheme, and the stringent scheme—all have very similar national benefits. Only the lenient scheme generated substantially fewer benefits. The similarity of national benefits among the four control schemes has implications for the distribution of benefits by income group: the income distributional impact of the benefits of control are also likely to be similar. The average air quality benefits for the current scheme (measured as pollutant concentration improvements) were found to be approximately the same for households in all income groups. Thus, the distribution of benefits of the alternative schemes, with the possible exception of the lenient scheme, should be roughly proportional because of the similarity of the total benefits for these other schemes.

The distribution of benefits under the lenient scheme should also be similar to the current scheme, but for a different reason. Recall that the average benefit for an income group is calculated as a weighted sum of the benefits obtained in each of the 487 geographic areas, with the weights being the fractions of households in that income group living in each area. If an alternative scheme simply lowers benefits by the same percentage in each geographic area, the benefit distribution by income group will remain the same, although of course the general level will fall. That is precisely what should happen under the lenient strategy. The lenient strategy changes the new car emission rates for each model year. But the other determinants of total automobile emissions—deterioration rates for older cars, the number of vehicle miles of travel, and the distribution of cars by car age—are not changed. As a result, benefits in each urban area should simply be decreased by the same percentage in each area.

This formula for calculating average benefits by income group suggests that there may be some differences in income distributional patterns among the four other schemes although, as mentioned above, the similarity of total benefits constrains these differences to be relatively minor variations. For example, the I-M program required under the stringent scheme will reduce emissions from older cars more than from newer cars. Since geographic areas differ in the age profile of cars driven, if certain income groups are systematically concentrated in areas with particular car age profiles, the distribution of benefits under the stringent scheme may be somewhat different than under the current scheme. Predictions of this sort could also be made for the two multicar strategies.

Table 8–8 shows the benefits by income group for each scheme. As predicted, the five strategies have virtually the same distributional patterns. This similarity is evident in the pro-poor ratios (the ratio of benefits for the lowest income group to benefits for the highest income group), which are listed in

Table 8-8. National Annual CO Concentration Benefits by Income
Group of Alternative Strategies in 1990

Strategy	<3	3-5	5-7	7-10	10-15	15-25	25+	Pro-Poor Ratio
				Income Group				Pro-Poor
Lenient	5.5	5.5	5.6	5.1	4.9	5.3	5.6	.98
Two car	10.1	10.2	10.2	9.5	9.1	9.9	10.7	.95
Three car	10.2	10.2	10.3	9.6	9.2	10.0	10.8	.95
Current	11.0	11.0	11.1	10.3	9.9	10.7	11.4	.96
Stringent	11.1	11.1	11.2	10.4	10.0	10.7	11.5	.97

the last column and shown in a bar graph in Fig. 8-5. Pro-poor ratios range from
0.95 for the two car and three car strategies to 0.98 for the lenient strategy.
While the multicar strategies are slightly less desirable than the current scheme
on equity grounds, the differences are negligible.

The results on the overall equity impacts of alternative schemes,
taking into account both the cost and the benefit impacts, appear to reinforce
the economic efficiency results. The two car strategy is best on both accounts: it
has the least regressive income-distributional consequences (along with the three
car scheme) and the most favorable cost-effectiveness consequences. Under the
two car scheme the bulk of the benefits of the current auto emission control
program are retained, while the total costs of emission control, and especially
the costs borne by households in lower income groups, are reduced.

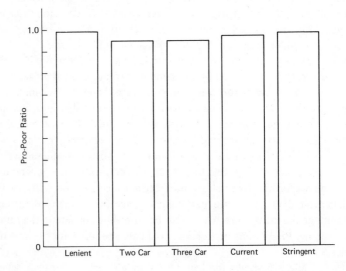

Figure 8-5. National Pro-poor Ratios for Alternative Strategies in
1990

NOTES TO CHAPTER EIGHT

1. National Academy of Sciences, *A Report by the Committee on Motor Vehicle Emissions* (Washington, D.C.: National Academy of Sciences, February 1973), p. 91.
2. U.S. Environmental Protection Agency, *Supplement Number 2 for Compilation of Air Pollutant Emission Factors,* Report AP-42, 2nd ed. (Washington, D.C.: U.S. Government Printing Office, September 1973), pp. 3.1.2-2 and 3.1.2-6.
3. 36 FED. REG. 8187 (1971).

Chapter Nine

Conclusions

Federally mandated controls on automobile emissions constitute the major government effort to improve air quality in the United States. This study has investigated the economic impacts of the current federal program and a number of plausible alternatives, concentrating on the impacts of auto emission controls for households in various income and geographical groups. Despite the uncertainties and difficulties of this enterprise, several important conclusions about federal automobile emission control emerge from this study. These conclusions fall into three major areas: national or aggregate benefits and costs, income-distributional impacts, and geographic variations. Comparing the current scheme with several alternatives resulted in a judgement that an alternative automobile emission control scheme would be economically superior, both from the standpoints of economic efficiency and equity.

This study's conclusions about the overall national impacts of the current auto emission control strategy are essentially the same as those reported in several other studies: the aggregate cost of the program will be quite large, but it will result in very large reductions in the level of air pollution experienced by American households. The annual costs of the auto emission control strategy mandated by the 1970 Clean Air Act amendments are estimated to be about four billion dollars in 1990, or $62 per household. These expenditures would reduce the average pollutant concentrations American households experience by 81 percent for carbon monoxide, 68 percent for photochemical oxidants, and 48 percent for nitrogen oxides.

This study evaluated four alternative auto emission control programs. The selection of alternatives was not meant to be exhaustive but rather to allow comparison of the current scheme with several plausible schemes which have been suggested as replacements. All but one of the alternatives involve lower total cost than the current scheme. A lenient strategy, in which the current emissions standards are replaced by the more lenient 1973 model year standards, costs about $1.8 billion, less than one-third as much as the current strategy. A two car scheme—in which the current new car emission standards only apply to cars in the 43 most polluted SMSAs while the lenient 1973 standards apply elsewhere—has an annual cost of $2.6 billion, roughly two-thirds of the current scheme. Adding an inspection-maintenance program to the two car scheme for the 19 most polluted urban areas, thereby creating a three car scheme, adds another $100 million to the cost of control. The only alternative considered in this study which is more costly than the current scheme adds a nationwide inspection-maintenance system to the current scheme; it has an annual cost of $4.5 billion.

These less costly alternatives produce smaller reductions in pollution concentration than the current scheme. But when the alternatives are judged from the standpoint of their cost-effectiveness, one scheme, the two car strategy, emerges as clearly superior. The two car scheme achieves almost all of the benefits of the current auto emission control program, 95 percent according to the results of this study, while saving over one-third of the control costs. Indeed, the three strategies that are more costly than the two car scheme—the three car scheme, the current scheme, and the addition of a national inspection-maintenance program—all result in very small increases in national air quality benefits over those achieved by the two car scheme.

A two car strategy also has far better equity consequences than either the current strategy or the other alternatives because under it the costs fall less heavily on the poor. This superiority is only in comparison to the other schemes considered in this study. Automobile emission control as a general approach to cleaning the air has disturbing income redistributional consequences, at least on the cost side. While the absolute dollar burden is lower for low income households, the percentage of household income taken by control costs is substantially greater for low than for high income households. In economist's jargon, the cost pattern for automobile emission control is *regressive.* All the control schemes, including the two car scheme, exhibit this regressive tendency.

For the current scheme, the average household earning less than $3,000 per year pays over three times as large a percentage of its income for automobile emission control as the average household earning more than $25,000 per year. This ratio of cost burdens is approximately the same for the lenient scheme and was somewhat higher for the most stringent scheme which incorporates a national inspection-maintenance program. The two car and the three car schemes have the *least regressive* cost burden pattern. The ratio of low income burden to high income burden declines to only 2.4 when the two car scheme is instituted. The ratio is the same for the three car scheme as for the two car scheme because the bulk of the costs are common to the two schemes; the additional costs for inspection-maintenance in the most polluted areas are only 4 percent of the costs of the three car scheme.

The equity effects on the benefit side do not differ for the five auto emission control schemes considered in this study. Although of course the level of benefits differs, the *variation* in benefits for households in different income groups is virtually identical for each scheme. The variation in physical benefits— measured as improvements in air pollutant concentrations—is nil under all schemes. This result was not expected. It was expected that the air quality benefits of auto emission control would be distributed in a markedly pro-poor manner, because lower income groups are concentrated in large central city areas where air quality benefits are very large. But lower income groups also make up a larger than average percentage of households in nonurban areas, where air quality benefits from emission control are very small. These two factors tend to cancel each other out.

Since the cost impact is expressed as a percentage of household income, the benefits should be expressed in dollar terms so that an analogous calculation of benefits as a percentage of household income could be made. However, the dollar values that households place on these physical benefits are unknown. If dollar values were known, the distribution of benefits measured by dollar benefits as a percentage of income would probably favor the lower income groups. But the extent of pro-poorness may be small because households in lower income groups may place low dollar values on air quality improvements.

The income distributional conclusions for all the automobile emission control schemes considered in this study rest on calculations that *understate* the costs and *overstate* the benefits to lower income households. Several factors may make the costs of auto emission control fall more heavily on the poor and the benefits accrue proportionately less to the poor than the calculations in this study indicate. For example, the cost distribution would be more regressive if the maintenance and fuel penalties increase with a car's age, rather than remaining stable for the life of the car as these calculations assume. The benefit figures assume that households receive benefits only where they live, be it central city or suburb. If the central city air quality benefits higher income suburban residents experienced from working or shopping in core areas were included, the benefit pattern would be altered to favor the rich.

The explanation for the superiority of the two car strategy over the current scheme on both economic efficiency and equity grounds lies in the wide geographic variation in costs and benefits under the current nationally uniform strategy of automobile emission control. The 1970 Clean Air Act Amendments mandated stringent emission requirements on cars sold *everywhere* in the United States. But while the standards are the same everywhere, the costs and benefits vary a great deal. The two car scheme obtains superior results by exploiting these geographic variations. The two car scheme is more cost-effective than the current scheme because it concentrates the costly stringent auto emission controls in large central cities, where pollution is high and large reductions in concentrations are achieved and where per household costs are low. The burden of control costs falls less heavily on the poor under the two car scheme because lower

income households are less highly represented in the large urban areas that require the more stringent controls.

The conclusions of this study regarding systematic geographic variations in economic impacts under the current scheme are based on a division of
households into five population size groups. Households in the 243 SMSAs in
the continental United States were divided into four population size groups ranging from SMSAs with more than four million persons to SMSAs with less than
300,000 inhabitants. The fifth group comprises the non-SMSA households.
Separate estimates were prepared for central city and suburban households in
the SMSA population size groups.

The first conclusion which emerged from the cost and benefit break-
downs by population size group was that the current auto emission control
scheme results in undesirable impacts for households in the suburban areas of
American SMSAs. Suburban residents in all population size groups on average
obtain low benefits and pay high costs. Moreover, these costs are borne quite
heavily by lower income groups.

Nonurban households also fare poorly under the current auto emission control scheme. Because of generally lower income levels compared to
suburban residents, and thus lower levels of car ownership, non-SMSA households pay considerably lower average costs than suburban residents, $60 per
household compared to $73 per household. But these costs fall even more heav-
ily on the poor in rural areas than they do in suburban areas. If suburban and
nonurban households are combined, approximately two-thirds of American
households are found to do badly under the current automobile emmision control program, obtaining small benefits while paying large costs which are borne
heavily by lower income households.

Central city residents do much better, but there are wide variations
by population size group. Using any measuring rod, households living in the
central cities of large urban areas obtain the best results. Households in the largest cities receive the largest benefits, pay the smallest costs, bear the least regressive cost pattern, and (when their suburban counterparts are included) have the
second most pro-poor benefit pattern. The variation in average benefits is particularly dramatic. City residents in SMSAs with more than four million people
on average receive a 69.6 ppm improvement in CO concentrations, compared
to 17.2 ppm in the one to four million group, 8.0 ppm in the 300,000 to one
million group, and 5.7 ppm in the less than 300,000 group. The variation in
average cost is less dramatic but still quite marked. Average costs for city residents range from $70 per household for SMSAs with more than four million
people to $62 for SMSAs with fewer than 300,000 persons.

These marked geographic variations in the cost and benefit impacts
of the current scheme should come as no surprise. The cost differences reflect
differences in automobile ownership patterns which in turn result from differences in available transport alternatives and urban spatial structure. Because

suburban areas generally have more dispersed origins and destinations, inferior public transportation, and higher average incomes, suburban households have higher levels of automobile ownership than their central city counterparts. Even low income households in suburban areas have high auto ownership rates, which accounts for the more regressive cost burden pattern in suburban areas. The situation is similar for nonurban households, except that, as noted above, the generally lower incomes for rural families imply lower average auto ownership rates and lower average cost burdens.

The seeming paradox of lower costs and greater benefits for central city residents is easily explained. Since air quality in suburban areas and non-urban areas is generally very good without the stringent auto emission standards, even very large percentage reductions in auto emissions can only yield modest improvements in the concentration exposure for suburban residents. In contrast, central city areas, with their concentrated patterns of economic activity and automobile travel, have high uncontrolled pollutant concentrations. Central city areas therefore stand to gain a great deal from strict auto emission controls.

These expectations about central city car ownership levels and concentrated activity patterns are primarily for the central cities in large urban areas. In small and medium sized central cities the situation is closer to the suburban and nonurban pattern: car ownership levels are relatively high, even among low income groups, and uncontrolled pollutant concentrations are relatively low. It is in the large central cities that households are able to achieve tolerable mobility without owning a car, because origins and destinations are likely to be close together and reachable by relatively good public transportation systems. These large central cities have the concentrated patterns of economic activity which generate large uncontrolled pollutant concentrations, and thus large air quality benefits from auto emission control.

The geographic variations under the current auto emission control program generate *regional* differences in cost and benefit impacts. Urban areas were divided into four regions—Northeast, North Central, South, and West. Households in the Northeast region stand to gain considerably more than any other region from the emission control strategy. Cities in the Northeast have the smallest average cost, the highest average benefits, the least regressive cost burden pattern, and the second most pro-poor distribution of benefits. The differences in average impacts are particularly large between the Northeast and the other regions. Average costs in the Northeast are barely 60 percent of those in other regions and average CO concentration benefits are over three times those of any other region.

In contrast, cities in the West and South fare badly under the auto emission control scheme. Cities in those regions bear quite high average costs and obtain relatively low average benefits. Costs are more burdensome to the poor and the benefits patterns are considerably less pro-poor in those regions. The North Central cities have cost and benefit impacts which fall in the middle ground between patterns in the Northeast and in the South and West.

These regional differences follow from differences in the urban spatial structure of cities in the various regions of the country. The older cities of the Northeast have compact residential and employment patterns which permit residents to avoid cost burdens by maintaining low levels of automobile ownership. These compact development patterns also result in high uncontrolled air pollution levels and thus in high benefits from auto emission control. The newer cities which predominate in the South and West have more dispersed residential and employment patterns and greater dependence on the automobile for mobility. Households in these newer, less dense cities pay high control costs while obtaining modest control benefits.

The original motivation for this chapter's discussion of geographic variations in costs and benefits under the current auto emission control program was to explain the superiority of the two car scheme. The two car scheme achieves more desirable efficiency and equity effects by mitigating the large variations in impacts for households in different population size groups and regions which occur under the current scheme. Under the two car scheme, the large urban areas continue to do well, since households there are still subject to the current emission control standards. But the more lenient standards permitted in smaller urban and rural areas significantly decrease the costs while not greatly affecting the benefits of control accruing to households in those areas. The situation is the same for regional variations: households in the large, older cities of the Northeast continue to do well under the two car scheme while households in the smaller and newer cities which predominate in the South and West are allowed to decrease the control cost burdens they bear with little sacrifice in air quality benefits.

The two car scheme's ability to reduce these geographic variations is an important result in its own right. Geographic variations do not fall under the traditional headings of economic efficiency and equity, but they are nonetheless important. Indeed, geographic differences in government programs are common and can be the source of considerable political controversy. For example, the recent concern for energy self-sufficiency has dramatized the greater dependence of the Northeast on foreign oil and thus the greater likely burden for Northeast households if policies to curtail foreign supplies or increase foreign oil prices are instituted. However, regional differences are not usually well documented, and general impressions about regional advantages and disadvantages may be inaccurate. Two of the objectives of this study were to document the regional variations in cost and benefit impacts of the current auto emission control program and determine if an alternative scheme would have preferable regional impacts.

The overall superiority of the two car strategy can be summarized succinctly: a two car strategy achieves most of the benefits of the currently mandated scheme with substantially lower total costs, a cost pattern that falls less heavily on the poor, and less geographic variation in economic impacts. Alternatives which retain the national character of the current strategy but make the standards either more lenient or more stringent appear economically inferior to this geographically variable alternative.

The two car strategy, however, is not without its limitations. One of the objectives of stringent automobile emission control is to permit highly polluted urban areas to achieve federally mandated ambient air quality standards for the three auto pollutants, carbon monoxide, nitrogen oxides, and photochemical oxidants. According to the results of this study, 19 central cities do not achieve the ambient standards for one or more of these pollutants in 1990 under the two car scheme. Some additional strategy would need to be instituted to reduce concentration levels in these 19 cities to the federal standards.

This study investigated a scheme to improve air quality in these 19 cities further by adding an inspection-maintenance scheme to the current emission standards. This three car scheme was judged not to be cost-effective in comparison to the two car scheme, since the additional air quality benefits in these 19 areas are quite modest in comparison to the extra costs. But since air quality benefits are not given a dollar value in this study, it is impossible to judge whether these additional benefits are worth the costs from an economic efficiency standpoint. It does seem likely, however, that some other scheme to reduce air pollutant concentrations in these areas, such as controlling stationary source emissions or regulating auto travel, would be more cost-effective than this three car scheme.

This limitation of the two car scheme points out some more general limits of this study. This study has broadened the scope of economic evaluation of automobile emission control beyond aggregate costs and benefits to provide estimates of the *differential* impacts of the current policy and several plausible alternatives on households in different income groups and different geographic areas. These impacts were seen to vary enormously for the current program, and these variations pointed to the economic superiority of a two car scheme. All these conclusions bear on the wisdom of the current automobile emission control strategy and its possible replacement by another scheme.

But a complete evaluation would include much more. A public policy evaluation forming the basis for firm conclusions in this area might expand the analysis along two dimensions: the range of alternative possible actions considered could be extended; and the number of impacts considered could be increased.

The automobile emission control program is only one program among many possible means of improving air quality. Ideally, government decision makers deciding on the wisdom of retaining or modifying the current strategy would have information on the costs and benefits of a wide range of alternative programs. For example, one option would be to reduce stationary source emissions to achieve improved air quality. Taking a somewhat broader view, one would want to know the impacts of other public health measures, such as immunization programs, subsidies for physicians services, and the like. The list could be extended still further to include virtually all government programs since they are all in some sense alternatives to the current policy of controlling auto emissions.

Expanding the list of impacts might involve analyzing the legal, administrative, and political aspects of alternative strategies. For instance, the two car strategy may have to surmount some legal, institutional, or political obstacles. Because these noneconomic factors were excluded, the analysis in this study does not allow one to conclude that one strategy is superior to the others in a final sense.

The results and conclusions of this study are meant to be a contribution to a comprehensive evaluation of the federal automobile emission control program. Studies analyzing other programs and taking other perspectives should complement the results and conclusions reached in this study.

Appendixes

Appendix A

Valuing Air Quality Benefits

The objectives of this appendix are to discuss the problems that arise in trying to quantify the dollar value of improvements in air quality due to automobile emission controls and to present a systematic review of the available evidence on the health hazards and other damages caused by the automobile related pollutants.

DOLLAR QUANTIFICATION OF AIR POLLUTION BENEFITS

This study predicts air pollutant concentration benefits for household groups crossclassified by household income and geographic area. As mentioned at several points in the study, it would be desirable to place a dollar value on the concentration benefits each group receives. These dollar values would reflect the "willingness to pay" (i.e., the amount of money the households would be prepared to sacrifice) for these air quality benefits. Dollar values for benefits would permit calculation of net benefits (benefits minus costs) for each group and also permit a more accurate calculation of variations in benefits for households in different income groups and geographic areas.

An example may be useful to illustrate the advantages of having detailed dollar values for household groups and also to show the difficulties of actually estimating these values. This study found that national air quality benefits were roughly the same for all income groups. But does this imply that the dollar value of benefits is the same? Would the average household in each income group be willing to sacrifice the same dollar amount for these air quality benefits? There are several reasons for expecting systematic variations by income group in the dollar values. One reason follows from the possibility that the health and other damages prevented by a given reduction in air pollutant concentration may be quite different at different starting points. If, for example, damages from higher concentrations declined as concentration increased, households experiencing the same absolute improvements at different starting points would obtain quite different improvements in health and welfare. This situation is illustrated by the air pollution damage function in Figure A-1. The points X_0 and Y_0 indicate the uncontrolled air pollutant concentrations household X and household Y would experience, while X_1 and Y_1 indicate the exposure concentrations with auto emission controls. Both households receive the same absolute change, but the benefits of reduced damages for household X, B_X, are much smaller than the benefits for household Y, B_Y. If households in different income groups start out with different average exposures, the change in damages from the same air quality benefits may be quite different.

In addition, households in different income groups may place different dollar values on given reductions in *damages.* Low income groups may be more tolerant of the aesthetic disbenefits of air pollution or may be less concerned about the increased disease rates caused by higher concentrations. For those subsisting in urban or rural slums with little to eat and few possessions, the health, aesthetic, and other benefits of improved air quality may be of small consequence. Higher income households already living in more pleasant circumstances may be much more intolerant of the hazy skies and health hazards of highly polluted air. Translated into valuation terminology, these higher income households should be *willing to pay* considerably more than their lower income counterparts for the same air pollution damages.

No empirical studies have been done to measure variations in willingness to pay for detailed household groups. However, a number of studies have attempted to measure the total willingness to pay for air quality improvements experienced by all American households. While these studies ignore the problems of variations in valuation by income group or geographic area, they

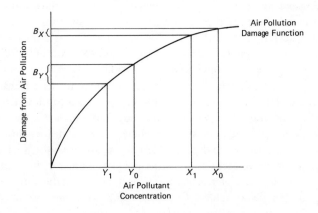

Figure A-1. Illustration of Variations in Air Pollution Benefits for a Given Change in Air Pollutant Concentration

illustrate a number of other difficulties involved in assessing dollar values for air quality benefits.

These air quality studies build on the tools developed by economists to measure the dollar value of benefits for large government projects, such as public education, flood control, or public highways. These government projects all provide benefits which are not sold in private markets. Goods sold in markets have market prices which can be used to measure aggregate willingness to pay. For example, the benefits of a scheme to provide more bananas for every American household could be calculated using the prevailing market price for bananas to measure the common valuation of all households for an additional banana.[a] Since private markets for public schools or highways, or air quality improvements, do not exist, some other means of calculating total willingness to pay must be found.

The standard economic approaches to valuing nonmarket goods are either to use the concept of opportunity cost or to identify market situations where the value placed on the nonmarket good can be estimated from analogous market transactions. The opportunity cost approach in essence requires answering one of two questions: "How much are other costs increased by air pollution?" or its equivalent, "How much would other costs be reduced if the quality of ambient air is improved?" This approach has been used in a number of studies to evaluate the costs of air pollution to plants, materials, or human health. For example, Lave and Seskin used statistical estimates of the effect of air pollution on morbidity and mortality rates to compute increases in health costs and decreases in earning capacity.[1] In addition to the earnings lost from illness and premature death, Lave and Seskin included the payments for hospitals, nursing home care, services of physicians, and services of other health professionals. They concluded that $2.08 billion would be saved each year by a 50 percent reduction in air pollution levels in major urban areas. In dollar terms, this total for increased health costs and decreased earning capacity should represent the amount society would be willing to pay to avoid these adverse health effects.

The second major technique used to value improvements in air quality infers willingness to pay from analyses of the housing market. When families decide to buy a house or rent an apartment, they consider both the characteristics of the particular housing unit and those of the surrounding neighborhood, which include air pollutant concentration levels. If housing consumers value cleaner air, they presumably would pay more for a unit located in an area with good air quality than for an otherwise identical unit located in an area with poor air quality. The value housing consumers place on cleaner air can

---

[a]Of course there are still several problems in inferring aggregate willingness to pay even when market prices exist. Is the prevailing price an equilibrium one? What about the possibilities of distortions as a result of market imperfections? Do prices vary for different urban areas? What is the proper price for nonmarginal changes? But these problems are minor compared to those involved when no market prices exist.

then be inferred from the difference in the market prices for comparable units in the neighborhoods with good and poor air quality.

Several studies have used this approach to obtain quantitative estimates of the influence of air pollution on property values. One of the first studies, by Ridker and Henning, used census tract data for St. Louis and the technique of multiple regression to estimate the effect of sulfation levels (SO_3) on property values.[2] Ridker and Henning found that sulfation was an important variable in explaining housing prices, but that the precise magnitude of the estimated effect was quite sensitive to the statistical specification assumed for the analysis. What they regarded as their best estimate implied that a 0.25 mg/100 cm^2/day change in SO_3, *ceteris paribus,* would increase the value of an owner occupied single family housing unit by $245. This result can be interpreted as indicating St. Louis homeowners would be willing to pay at least $245 to live in areas where the SO_3 level is 0.25 mg/100 cm^2/day better.

These two approaches to valuing willingness to pay—the opportunity cost approach and the housing market approach—may be more complementary than competing ones. Most opportunity cost studies measure the costs of damages to human health, materials, or vegetation. The housing value studies are more concerned with the aesthetic harms from higher concentrations, since it is these aesthetic effects which households are most likely to perceive when they evaluate different neighborhoods.

Although several other studies have used one of these two approaches to valuing air pollution improvements, these studies are of limited value in evaluating the dollar benefits of automobile emission reductions because the automobile pollutants typically are not included in the studies.[b] For example, Lave and Seskin's estimate of health damages is based on studies which use either particulates or sulfates to measure air pollution. They argue that these variables represent the effects of a larger number of pollutants, including those emitted in large quantities by automobiles. Similarly, all of the published housing market studies use particulates or sulfur oxides to measure air pollution.[c] The emphasis on nonautomobile pollutants in these quantitative studies partly reflects on a priori belief that these substances are more harmful and provide better aggregate estimates of the seriousness of air pollution. But an even more compelling reason is that there are many more monitoring stations for sulfur oxides and particulates which have been in operation for a longer period of time.

---

[b]After this study was completed, the National Academy of Sciences conducted a study, in which the author participated, of the national costs and benefits of the current automobile emission standards under contract from the U.S. Senate Public Works Committee. The Academy study concluded that the most reasonable estimate of the total annual benefits of controlling auto emissions was $5 billion. This figure is based on property value studies, wage rate studies, and health and welfare studies.[3]

[c]The recent National Academy of Sciences study included an empirical analysis, performed by the author and Robert MacDonald, of the influence of automobile pollutant concentrations on housing prices in the Boston and Los Angeles urban areas.[4]

This study did not develop dollar values for the automobile pollutants since, as mentioned above, this study actually requires much more detailed dollar values for household groups. However, the next section of this appendix provides systematic summaries of the existing evidence on the adverse effects of the three automobile pollutants. While this evidence is incomplete, it should give the reader a sense of the types of improvement in health and welfare American households will experience because of the automobile emission control program.

AUTOMOBILE AIR POLLUTION DAMAGE STUDIES

This section presents brief summaries of the available evidence on the damages caused by carbon monoxide (CO), nitrogen oxides (NO_x), and photochemical oxidants (O_x). The studies pertaining to the health effects of the air pollutants are reviewed first. This survey of the health effects literature is then followed by a discussion of the aesthetic effects of the air pollutants and brief summaries of the major studies that have identified air pollutant damages to materials and vegetation.

Health Effects

Most public concern about automobile emissions derives from their alleged effects on disease and death rates. Although the evidence on these adverse health effects is far from complete or certain, there is a growing body of medical literature which indicates that the health of urban residents would be improved and the lives of urban residents lengthened if automobile emissions were reduced/ Before considering individual studies of the effects of specific pollutants on health, it is useful to review the methodology used by medical researchers. Some understanding of the methodology is needed to appreciate why research on the effects of air pollutants on health is so difficult and why the available evidence on these effects remains so uncertain.

Two complementary approaches have been used to evaluate the health effects of exposure to concentrations of various air pollutants: (1) clinical laboratory studies and (2) statistical epidemiological studies. Clinical studies, which identify the physiological effects of pollutant inhalation on various body systems, are especially useful in *identifying* diseases which may be aggrevated by pollutant concentrations. These clinical studies cannot establish a causal relationship between pollutant concentration and human death or disease rates, because studies that use human subjects cannot employ doses that are large enough to risk death or disease. In addition, clinical researchers have generally limited their studies of both animals and humans to the effects of short term exposure to high concentration.

Statistical studies have a greater potential to quantify the long term health effects of air pollutants, but they too are fraught with difficulty. These epidemiological studies attempt to determine the relationships between morbidity and mortality rates and long term exposure to various pollutants by comparing death and sickness rates for populations in different geographic areas who are therefore exposed to different concentrations of particular pollutants.

Unfortunately, existing epidemiological studies are of limited use in inferring causal relationship between pollutant concentrations and human health. There are virtually no usable epidemiological studies for some pollutants, and the analyses of the well-studied pollutants often contain serious methodological flaws. The most serious problem is the lack of data on other factors which influence death and disease rates. In addition to air pollution, human health is presumably affected by diet, heredity, age, smoking habits, and general physical and mental well-being. Without information on other factors that affect health, it is difficult to make inferences of the health effects of exposure to particular pollutants. If one or more of these other variables is correlated with pollution levels, simple correlations between air pollution levels and health effects could be spurious.

Some studies which claim to assess the effect of particular air pollutants on mortality and morbidity make no effort to control for these other factors. Indeed, some studies do not even measure air pollutant concentrations, but simply infer the health effects of air pollution from differences between urban and rural death and disease rates. Many of the existing studies are plagued by additional statistical problems, such as a small sample size, which precludes reliable statistical results, or inadequate statistical techniques.

Despite these inadequacies, the existing health effects information does allow some inferences to be made about the probable health effects of the automobile air pollutants. The available evidence is summarized below.

Carbon Monoxide.   Carbon monoxide has long been known to cause death at very high concentrations. But until recently there was no evidence to suggest that it was harmful at the low to moderate concentrations found in urban airsheds.

Clinical studies and theoretical medical research provide a detailed description of the physiological effects of breathing carbon monoxide and of its influence on body systems.[5] Carbon monoxide is absorbed into the lungs and reacts with the blood component hemoglobin (the carrier of oxygen) in the blood to form carboxyhemoglobin (COHb). It effects the body's oxygen supply in two ways. First, since the bond between carbon monoxide and hemoglobin is roughly 200 times as strong as that between hemoglobin and oxygen, COHb reduces the oxygen carrying capacity of the blood. Second, COHb interfers with the oxygen releasing mechansim of hemoglobin and impairs the release of oxygen in body tissues.

The quantitative relationship between the duration of exposure to various carbon monoxide concentrations and blood levels of COHb has also been studied. Clinical research indicates a fairly consistent relation between CO concentration and the increased percentage of COHb in the blood. This relation depends on the duration of exposure to CO and the breathing rate. A higher than normal breathing rate, as would take place during exercise, would speed the approach to the equilibrium level of COHb.

Whereas the physiological effects of CO exposure are well established, their implications for health are not. Several clinical studies suggest that increases in COHb have a harmful effect on the central nervous and cardiovascular systems. Central nervous system effects observed at low COHb levels (2–10 percent) include decreases in visual discrimination, vigilance, and general performance.[6] These studies require subjects to perform various tasks (such as indicating whether the intervals between sounds are the same or different) before and after their COHb levels are increased. Some of these studies, particularly those showing abnormal responses at very low (2 percent) COHb levels, are under question, and a few opposite results have been reported for low levels of COHb.[7]

Since the split second reactions required to drive safely on high speed highways may be dulled by higher CO levels, it has been argued that some fraction of traffic accidents can be attributed to the central nervous system effects of COHb.[8] This conclusion has not been established by any existing study. However, a few studies have reported indirect evidence of a link between COHb concentration in the blood and automobile accidents. Chovin measured COHb levels in three groups of Parisians—drivers involved in automobile accidents, workers exposed to high occupational levels of CO, and individuals suspected to having been exposed accidentally to high CO levels—and found COHb levels highest in the drivers. Ury found an association between traffic accidents and concentrations of photochemical oxidants in Los Angeles, and speculated that part of the association may be due to the presence of CO. Clayton et al analyzed blood samples of drivers and pedestrians and found a higher COHb level in the drivers.[9]

The most promising technique for considering the relationship between elevated COHb levels and traffic accidents would be an epidemiological study designed to test statistically for CO-associated increases in traffic accidents. To isolate the independent effect of CO concentrations, it would be necessary to account for fluctuations in traffic, visibility, vehicle conditions, and other driver conditions (particularly drinking). The substantial data and statistical problems encountered in carrying out a study of this kind have discouraged researchers from attempting this type of study.

High CO concentrations may also adversely affect the heart and circulatory system, particularly for persons with heart disease. These harmful cardiovascular effects have been studied in clinical experiments, mostly with animal subjects. Degenerative cardiovascular changes, possibly increasing the proclivity of the animals to develop atherosclerosis, were observed in rabbits whose COHb levels were raised to around 20 percent.[10] Human exposure studies have focused on subjects with pulmonary emphysema and coronary heart disease. Experimental exposures to CO resulting in COHb levels of around 5 percent have been associated with a number of cardiovascular changes that impose greater burdens on the heart and circulatory systems of members of this particularly vulnerable group of people.[11]

Two recent studies suggest that even lower levels of CO aggravate symptoms in patients with heart disease.[12] Both studies measured the time to the onset of anginal pain for ten patients exposed first to pure air and then to air with low levels of CO. In the first study exposures to 50 ppm CO and 100 ppm CO for four hours produced mean COHb levels of 2.9 percent and 4.5 percent in contrast to a mean control value of 1.3 percent. Nine of the ten patients developed anginal pain earlier during exercise (on a treadmill) after breathing CO than after breathing plain air. The findings of the second study, which used bicycle ergometer exercise, were similar. The mean COHb level after 50 ppm CO exposure for two hours was 2.68 percent in contrast to a control value of slightly less than 1 percent. Anginal pain occurred earlier after CO exposure than after air breathing in all ten patients. These clinical studies suggest that continued exposure to higher levels of CO may result in cardiac strain, and perhaps early deaths, for those with pre-existing cardiovascular disease.

Epidemiological studies have also been used to consider whether or not fatality rates for persons with heart disease are higher at high CO concentrations. For example, Cohen et al. compared the fatality rates of two groups of Los Angeles residents hospitalized for myocardial infarction (heart disease) in 1958, those hospitalized in areas of "high" pollution and those hospitalized in areas of "low" pollution.[13] They found that the "high" pollution group had higher fatality rates, but only when the general level of CO concentration was high. The finding that the difference was evident only when CO concentration was high supports the hypothesis that differences in CO levels, rather than other systematic differences between the two groups such as differences in the quality of health care, were responsible for the difference in fatality rates.

Hexter and Goldsmith concluded from a study of total daily deaths in Los Angeles that higher CO levels are associated with excess mortality.[14] The study included seasonal indexes and temperature as additional independent variables. While the findings of the Hexter and Goldsmith study are suggestive, they are difficult to interpret. Increased daily CO levels may only hasten death in those with serious pre-existing health problems. If fluctuations in daily CO levels were reduced, the lifespan of very ill persons might be increased by only a few days. Also, the Hexter-Goldsmith study does not clarify the possible long term adverse effects of exposure to higher ambient CO levels. Even the short term effects identified by this study are questionable since the levels of other pollutants and other environmental variables were not included as explanatory variables. If these excluded variables affect daily death rates and are positively correlated with CO levels, the estimated effect of CO levels on daily death may be spurious or at least exaggerated.

A far greater number of persons may suffer minor health problems or discomfort, such as occasional headaches from lowered blood oxygen levels, when CO concentrations are high. No clinical and epidemiological studies have been done to assess the independent influence of CO on these minor illnesses. Cassell et al. did analyze the effects of environmental influences on colds, coughs, headaches, and eye irritation.[15] However, since this study focused on the reinforcing influences of several environmental factors, it provides no estimate of the independent influence of CO concentration on these minor illnesses.

In summary, clinical and epidemiological evidence on the health effects of CO concentrations provides considerable basis for public concern. There is considerable evidence that high levels of CO impair the heart and circulatory system and increase the risk of heart attack for persons with cardiovascular diseases. Reactions dulled by CO concentrations may also be a contributing cause of highway injuries and deaths. In addition, high CO concentrations may be responsible for headaches and other minor ailments afflicting a much larger number of people.

Nitrogen Oxides. Nitrogen dioxide (NO_2) is the only oxide of nitrogen believed to have adverse health effects at ambient air concentrations. Clinical studies on animals indicate that the major effect of NO_2 is on the respiratory system. The basic pathological response is an inflamation of the lungs, accompanied by various chemical and cellular changes in lung tissue, which include damage to lung proteins, rupture of mast cells, and development of pre-emphysematous lesions.[16] Several effects on other body systems have been observed in animals exposed to NO_2: tissue damage to the kidney, liver and heart; weight loss; depressed voluntary running activity; and changes in the blood composition.[17]

The respiratory system effects of NO_2 seem to have two major health consequences, a greater incidence of chronic lung disease and a greater susceptibility to respiratory tract infection. A large number of studies suggest there is a connection between NO_2 concentrations and chronic lung disease, but the evidence is drawn primarily from animal experiments. These studies identify certain pathological changes in animals after exposure to NO_2 which appear similar to changes that occur in pulmonary disease in man.[18] A smaller number of studies of humans exposed to high concentrations of NO_2 for short periods of time report impaired pulmonary function and thus seem to confirm this relationship.[19] For obvious reasons, the effect of long term, low level exposure to NO_2 on human lung disease rates has not been clinically investigated.

The link between exposure to NO_2 and increased susceptibility to respiratory disease is suggested by animal studies using both short term and long term exposures. Animals exposed to NO_2 have shorter lives and are less able to clear infectious agents from their lungs. The major long term study, by Ehrlich and Henry, involved the exposure of groups of mice to quite low concentrations (0.5 ppm) of NO_2 for periods up to 12 months.[20] Ehrlich and Henry reported statistically significant increases in mortality after continuous exposure for three months and six months, and after 18 hour exposures for six months. No clinical study with human subjects has investigated this health effect.

The only major epidemiological studies of NO_2 health effects are based on data for children living and attending elementary school in four Chattanooga residential areas in 1969.[21] One area, located close to a large TNT plant, had high NO_2 and low particulate concentration; a second area had high particulate and low NO_2 exposure; and the two remaining areas served as "clean" controls. The TNT plant had been operating for about three years when the health data were collected. Three health effects were considered in the two studies which used this data: ventilatory (lung) function, the frequency of acute upper respiratory illness, and the frequency of lower respiratory illness.

The original study considered ventilatory function and acute respiratory illness, and the authors concluded that higher nitrogen dioxide concentrations did impair health. The study found that the lung capacity of children in the high NO_2 area was inferior to that of children in the control areas and that the families of these children had 18.8 percent more respiratory illness than families in the control area.[22] The retrospective study of lower respiratory illness indicated that both infants' (up to three years old) and school children's exposure to intermediate and high levels of NO_2 was associated with a significant increase in the frequency of acute bronchitis.[23] Illness rates for other lower respiratory diseases did not seem to be affected by NO_2 levels.

The Chattanooga study was designed to measure acute, relatively short term effects of greater NO_2 concentrations. While the Chattanooga researchers collected information on chronic disease among the study children, they made no attempt to evaluate the effects of differences in NO_2 levels on these chronic disease rates. Even so, the Chattanooga findings suggest that NO_2 concentrations are more likely to affect short term respiratory diseases than long term chronic disease rates, since the data indicate that the severity of upper respiratory disease (as measured by the presence of fever, home confinement, or visit to a physician) is no greater in areas of high NO_2 concentration than in the control areas. However, since the TNT plant had been open for only three years, the area really was not suitable for the analysis of these long term effects.

General Motors' researchers have sharply criticized the Chattanooga study and the federal government's use of it to set federal air quality standards for NO_x.[24] They have raised questions about nearly every aspect of the study's methodology: the techniques used to measure ambient NO_2, the classification of schools as high NO_2 and control, the medical significance of the observed differences in ventilatory function, the interpretation of the acute respiratory illness results, and the change in classification of the subareas in the retrospective study. Finally, they claim that health effects from increases in NO_2 concentrations cannot be inferred from the study because there is no evidence that exposure above an intermediate level gives rise to further impairment.

Researchers at the Environmental Protection Agency have answered these criticisms, although much of their defense relates to the general usefulness of epidemiological studies rather than General Motor's specific critique of the Chattanooga study.[25] EPA researchers also argue that the findings of the Chattanooga study are supported by a large number of laboratory studies which indicate that high nitrogen dioxide concentrations may increase the incidence of respiratory disease in man.

Photochemical Oxidants. Photochemical oxidants (O_x) include a large number of different substances, of which ozone is the largest single contributor. Many clinical studies have analyzed the damage to experimental animals resulting from high short term ozone concentrations, although the precise mechanism of ozone damage still is unknown. Short term exposure of experimental animals to ozone concentrations above 1 ppm have been shown to increase the incidence of respiratory tract infections, lower resistance to infection, reduce pulmonary function, cause visible damage to the lungs, and cause chemical and biochemical changes in nonrespiratory organs.[26] Less work has been done on the effects of long term exposure on animals, although several chronic respiratory diseases, such as bronchitis and emphysema, have been associated with long term exposure.[27] Additional effects on animals, including decreased fertility, increased mortality rates at birth, and increased stress response, have been observed after exposure to ambient photochemical smog in Los Angeles or irradiated automobile exhaust in laboratories.[28] However, because a large number of other pollutants are present in addition to the photochemical oxidants, these effects cannot be unambiguously attributed to high oxidant levels.

Very few studies have examined the damage to humans resulting from either ozone or total oxidant levels. There is no clinical or experimental evidence linking ambient oxidant levels to anything beyond discomfort, although the available evidence on impaired pulmonary functions is suggestive. A few experimental studies report that human pulmonary function is impaired by exposure to concentrations of oxidants ranging from 0.5 to 1 ppm for one to two hours.[29] In addition, industrial workers exposed to moderately high (0.3 ppm) to very high (3 ppm) levels of oxidants for periods ranging from one to one and one-half hours are reported to have unusually frequent nasal and throat irritations, distinct irritation of mucous membranes, coughing, irritation and exhaustion, and sleepiness.[30] In one case an extremely high concentration, far in excess of ambient concentrations (9 ppm), is reported to have caused respiratory disease.[31]

Several epidemiological studies have considered the health effects of high short term oxidant levels. It has been established that high oxidant levels are primarily responsible for the eye irritation experienced in Southern California urban areas, and a statistically significant relation was found between oxidant levels and high school athletic performance.[32] A study of symptoms in student nurses found that headache, eye discomfort, cough, and chest discomfort were related to daily maximum photochemical oxidant levels.[33] Most studies of the relationship between high oxidant levels and short term illness indicate that higher oxidant levels do not aggravate existing respiratory dieseases, such as asthma, bronchitis, and emphysema.[34] However, an early study of 137 patients with asthma demonstrated a significant difference in attack rates when daily maximum oxidant levels exceeded 0.25 ppm.[35] After controlling for the adverse effects of high temperatures, no study has demonstrated that oxidant concentrations make an independent contribution to daily death rates, although high oxidant levels have been linked to high automobile accident rates.[36]

There are no adequate studies of the long term effects of high oxidant concentrations on health. Several studies have explored these possible long term effects by comparing the incidence of chronic respiratory disease symptoms among Los Angeles residents to the incidence of symptoms among other California residents.[37] Since Los Angeles has both higher oxidant levels and a higher incidence of chronic respiratory disease than the rest of California, it is argued that this indicates that high oxidant levels have adverse long term effects. But, while this finding is suggestive, it is a weak test since many other peculiarities of Los Angeles may be responsible for its higher respiratory illness rates. Moreover, lung cancer mortality rates are lower in Los Angeles than in San Francisco or San Diego.[38]

In conclusion, there is abundant evidence that higher photochemical oxidant levels make life less pleasant, but there is little evidence yet that indicates that the effects are serious enough to increase disease or death rates. But until careful long term studies are done, the possibility that oxidants may cause more serious health effects cannot be discounted.

Aesthetic Effects

Air pollutant concentrations can reduce visibility, change the color of the sky, create unpleasant odors, and cause annoying eye irritation.

Reduction in visibility and changes in natural sky color are perhaps the most noticeable effects of air pollution.[39] Air pollution reduces visibility by absorbing light, thereby diminishing the contrast between the object being viewed and its background. Pollutants also scatter light in the space between the viewer and the object, further decreasing contrast.[40] Similarly, air pollution changes sky color by absorbing some color wave lengths more than others, producing ugly brown smog clouds. Although there have been several efforts to quantify the relationships between pollutant concentrations and these visual effects, none provide meaningful damage estimates.[41]

Other aesthetic damages are equally difficult to quantify. Most studies of air pollutant odors simply measure threshold perception levels and do not attempt to evaluate the discomfort caused by the odor. Yet obnoxious odors certainly have ill effects. In extreme cases they can cause temporary health effects such as loss of appetite, lower water consumption, impaired respiration, nausea, vomiting, and insomnia.[42] A California survey identified eye irritation as the most objectionable aspect of air pollution.[43] But like the other aesthetic effects, eye irritation damages from pollutant concentrations are difficult to evaluate. Individuals differ in their sensitivity and it is difficult to isolate specific eye irritants.

Carbon monoxide is a colorless, odorless, and tasteless gas with no known aesthetic effects and no effect on eye irritation.[44] Nitrogen dioxide (NO_2). the major atmospheric nitrogen oxide, is a reddish brown gas with a pungent odor.[45] It affects both atmospheric color and visibility by absorbing light. The absorption effect of NO_2 concentrations on visibility, however, appears much less important than the scattering effect of particulates,[46] and the coloration effects of NO_2 are reduced in the presence of substantial particulate concentration.[47]

The various contributors to total oxidants have different aesthetic effects although none have appreciable effects on visibility or coloration. Ozone has a pungent color at low concentration (0.02 ppm) which is instantly detectable by those with a normal sense of smell, although perception lasts less than a minute at these levels. At moderate concentrations (0.05 ppm), the ozone odor is considerably stronger and lasts longer (although still only a matter of minutes.)[48] Ozone is not an eye irritant.[49]

In contrast, the peroxyacetyl nitrates (PAN) do not have an odor, but they are among the major causes of eye irritation. The other major photochemical products that cause eye irritation are acroleon, formaldehyde, and peroxybensoyl nitrate (PB_zN).[50] Other products of photochemical reactions may also contribute to eye irritation.

Beyond the identification of some of the eye-irritating substances, relatively little has been done to quantify the relation between pollutant concentrations and eye irritation. An exception is a California study which reports statistically significant correlations between eye irritation and oxidant concentrations.[51] No significant correlations were found between eye irritation and concentrations of two other pollutants, NO_2 and particulates. This controlled experiment study was based on the responses of telephone company employees to conditions in filtered and nonfiltered rooms in downtown Los Angeles. The researchers also obtained a reasonably good correlation between the severity of eye irritation and the total oxidant level, with an eye irritation threshold at around 200 $\mu g/m^3$ (0.10 ppm).[52]

Materials Effects

Air pollution damages a wide range of materials including metals, building materials, painted surface, rubber, textiles, and textile dyes. Materials damage has been linked to concentrations of NO_x and O_x, but not CO.

However, it is often very difficult to distinguish the damages caused by individual pollutants from the effects of "weathering."[53]

Materials damage can take several forms. For example, air pollution corrodes metals, thereby requiring more frequent replacement and greater maintenance expenses for exposed metal parts. These damages can sometimes be avoided or be minimized, but the adjustments are invariably costly.

The major materials damage from NO_x is to textiles.[54] Nitrogen oxides damage textiles both by chemical reactions of NO_x with textile dyes and by weakening the textile fibers themselves. Damage to textile dyes appears to be the more severe problem.[55] The most detailed studies deal with the so-called "gas-fume fading" which occurs in rayon acetate.[56] This fading is uncommon today because research chemists have developed dyes for rayon acetate which resists the facing effects of NO_x. But these dyes are both more expensive and have poorer dyeing properties,[57] resulting in higher prices for clothes and other products which use rayon acetate.

Less information is available on the adverse effects of NO_x on fibers themselves. A Berkeley, California, field study reports a significantly lower mean breaking strength for combed cotton yarn samples exposed to unfiltered air than for those exposed to filtered air.[58] The Berkeley study does not unambiguously identify the pollutants responsible for the decrease in breaking strength, but the investigators concluded that NO_x was probably responsible. There is also some rather casual evidence that NO_x damages synthetic fibers. For example, an episode was reported in New York of runs in nylon stockings worn by women in the vicinity of a demolition project which created high levels of NO_2 and dust.[59]

Evidence linking NO_x and metal damage is also sparse, but nickel-brass corrosion has been traced to high nitrate concentrations in airborne dust.[60] This problem was first noticed in nickel-brass wire springs in telephone relays located in Los Angeles. Similar failures in relays and other nickel-brass components have been reported elsewhere in California and in several other states. Laboratory tests indicate that the stress-corrosion cracking of the wires only takes place when surface nitrate concentrations are about 2.1 mg/cm^2 and when relative humidity is about 50 percent.[61] In addition, Bell Laboratories has identified a second type of corrosion problem, which effects the nickel base of the palladium topped contacts of switches.[62] This second type of corrosion has been observed in a large number of cities including Cincinnati, Cleveland, Detroit, Los Angeles, New York, and Philadelphia.

These nickel corrosion problems are avoidable. Telephone company researchers have determined that stress-corrosion does not occur when zinc is eliminated from the nickel-brass alloy.[63] Thus, in newly manufactured wire spring relays a copper-nickel material is used. To protect existing nickel-brass relays, telephone companies in high nitrate areas have installed filtering systems and redesigned cooling systems to keep relative humidity below 50 percent. It is hard to estimate the increase in operating costs caused by these changes;

but they illustrate the nature of the increased costs resulting from NO_x-induced corrosion.

Ozone is the only photochemical oxidant whose materials effects have been analyzed. These studies indicate that ozone concentrations damage rubber, textile fibers, and textile dyes. Damage to rubber is caused by changes in the structure of polymers which occur when they are exposed to ozone.[64] These chemical changes lower the tensile strength, decrease the elasticity, and increase the brittleness of rubber. In tires, the major use of rubber, these changes cause cracking on the side wall part of the tire and in this way reduce tire life.[65] Similare cracking occurs on other exposed rubber products, such as conveyor belts, automobile rubber parts, and wire and cable. Ozone-sensitive rubbers accounted for 85 percent of the rubber production by weight in 1969.[66]

Tire manufacturers have developed synthetic rubber blends and anti-ozonant additives to protect tires and other exposed rubber products from ozone damage. But the addition of 1.5 percent antiozonants to automobile tires costs up to $0.50 per tire and does not entirely eliminate the ozone cracking problem.[67] Oils, gasoline, and other chemicals tend to extract antiozonants from rubber, leaving it again susceptible to ozone attack.[68] As tread wear on passenger tires improves and sidewall cracking becomes the limiting factor in tire life, the pressure to develop better antiozonants increases. These better antiozonants will likely cost more and will be reflected in higher tire prices.

Both textile fibers and the dyes used to color them are susceptible to ozone damage. Ozone attacks the cellulose in textile fibers and causes them to lose their strength and fluidity.[69] The effect of ozone on dyes was first discovered while testing NO_2-resistant dyes in Ames, Iowa, an area having a low NO_2 level but a fairly high ozone level.[71] When acetate fiber samples dyed with resistant dyes faded, scientists repeated the experiment in ozone controlled laboratories and discovered a large number of dyes were sensitive to ozone. Field studies in Chicago, Los Angeles, Phoenix, and Florida demonstrated that the extent of fading depended on both the concentration and duration of ozone exposure and relative humidity.[71]

It is hard to estimate the costs resulting from ozone damage to textile fibers. High ozone concentrations may shorten the life of products made with nylon and polyester fabrics, and thereby require consumers to replace clothing and other fabrics more often. The fading of ozone sensitive dyes can be counteracted by use of proper combinations of fabrics, dyes, and treatments. But these modifications are expensive, both because they require additional research expenses and because they use more costly materials. Consumers ultimately pay these higher costs in the form of higher textile product prices.

Plant Effects

Air pollution damages to plants have been studied extensively since they impose large economic costs on farmers and nurserymen. Both short term high concentration exposures and long term accumulations of the pollutants cause plant damage. High pollutant concentrations result in visibly damaged leaves and flowers which cannot be sold; lower concentrations for long periods of time might not produce any visible effects but do lower crop yields.

Households are affected by this damage by the higher prices they must pay for food and decorative plants and flowers. In addition, air pollution may reduce the lives of household shrubbery and other landscaping. Yet in spite of considerable research, the extent of plant damage due to air pollutants remains uncertain because of the difficulty of separating the independent effects of high pollutant concentrations from other climatic and environmental factors.

Carbon monoxide appears to have no detrimental effects on plants, even at very high ambient concentrations.[72] There is some evidence that high concentrations of carbon monoxide in the soil, which may result from high ambient concentrations, may limit plant growth by inhibiting the nitrogen fixing process of plant bacteria.[73] However, ambient concentrations of carbon monoxide probably cause only very minor plant damage.

Nitrogen oxides and oxidants have both been linked to substantial plant damage. Most laboratory studies of NO_x involve NO_2. The acute effects of NO_2 include cell collapse, patterns of localized cell death, or necrosis, and a waxy appearance.[74] The concentrations used in these laboratory studies are much higher than any ambient concentrations, but acute damage to plants has been observed in the vicinity of industrial sources such as nitric acid plants.[75]

Less information is available about chronic NO_x damage to plants. A diseased condition characterized by pigment changes in leaves and loss of leaves (leaf drop) has been observed in navel orange trees, tobacco, spinach, and soybean plants exposed to smog in the Los Angeles basin,[76] and yields of some crops exposed to moderate NO_x concentration for many months were reduced. These effects were attributed to NO_x because the symptoms of ozone damage were not observed. Heck et al. developed an empirical model to predict NO_x injury to vegetation.[77] These models and other empirical information have been used to classify plants according to their sensitivity to NO_x, but these data are not sufficient to make reliable predictions of plant injury at different NO_x concentrations.

The economic loss to farmers and nurserymen resulting from photochemical oxidants may be the largest of any of the air pollutants.[78] The effects of photochemical oxidants on crops in Los Angeles have been evident for a number of years. The primary oxidants responsible for injury to vegetables, ornamental flowers, and field crops have been identified tentatively, although the list may be incomplete. The two most studied oxidants are ozone and peroxyacetyl nitrates (PAN).

Ozone appears to be the most damaging photochemical oxidant.[79] Acute ozone-type injury is characterized by a spotting on leaves caused by death of cell tissue.[80] This injury is referred to as "weather fleck" in the tobacco growing areas along the eastern seaboard. In more advanced stages, these areas enlarge, coalesce, and eventually form lesions through the leaf. These lesions allow ozone to enter the leaf and cause further damage. Chronic injury from ozone concentrations is characterized by color changes as well as spotted patterns.[81] Clorosis, the loss or reduction in chlorophyll, is very common and produces a pale green or yellow color in leaves.

PAN-type injury is characterized by an undersurface glazing of the leaves caused by collapse in the spongy cells surrounding the air space in leaf openings.[82] PAN-type injuries to field crops have been reported in many parts of the U.S. and in foreign countries, although the damage occurs primarily in California and on the eastern seaboard.[83]

The evidence relating ozone, PAN, or other photochemical oxidants to reduced crop yields is primarily circumstantial. Yields have declined in Los Angeles and other areas for a number of years even when there has been no visible injury to the leaves.[84] Carnations exposed continuously over a two month period to low levels of ozone bore significantly fewer flowers and exhibited reduced height and stock length.[85] Still, no aggregate estimates of effects of oxidants on plant yields are available.

NOTES TO APPENDIX A

1. L.B. Lave and E.P. Seskin, "Air Pollution and Human Health," *Science* 169 (August 21, 1970): 723–733.
2. R.G. Ridker and J. Henning, "The Determinants of Residential Property Value with Special Reference to Air Pollution," *Review of Economics and Statistics* 49 (May 1967): 246–57.
3. National Academy of Sciences, A Report by the Coordinating Committee on Air Quality Studies, *Volume IV: The Costs and Benefits of Automobile Emission Control,* (Washington, D.C.: U.S. Government Printing Office, September 1, 1974).
4. Ibid., pp. 221–242.
5. U.S. Department of Health, Education and Welfare, *Air Quality Criteria for Carbon Monoxide* (Washington, D.C.: U.S. Government Printing Office, March 1970).
6. R.R. Beard and G.A. Wertheim, "Behavioral Impairment Associated with Small Doses of Carbon Monoxide," *American Journal of Public Health,* vol. 57 (November 1967), pp. 2011–12; R.R. Beard and N. Grandstaff, "CO Exposure and Cerebral Function" (Paper presented at the New York Academy of Sciences Conference on Biological Effects of Carbon Monoxide, New York City, January 12–14, 1970); M.M. Halperin et al., "The Time Course of the Effects of Carbon Monoxide on Visual Thresholds," *Journal of Physiology,* vol. 146(3) (June 11, 1959), pp. 583–93; R.S. McFarland et al., "The Effects of Carbon Monoxide and Altitude on Visual Thresholds," *Aviation Medicine,* vol. 15(6) (December 1944), pp. 381–94;

J.M. Shulte, "Effects of Mild Carbon Monoxide Intoxication," *Archives of Environmental Health,* vol. 27 (November 1973), pp. 524–30; and U.S. Department of Health, Education and Welfare, *Air Quality Criteria for Carbon Monoxide,* pp. 8–14 to 8–24.

7. U.S. Department of Health, Education and Welfare, *Air Quality Criteria for Carbon Monoxide,* pp. 8–18 to 8–19.

8. Ibid., pp. 8–24 to 8–34.

9. H. Ury, "Photochemical Air Pollution and Automobile Accidents in Los Angeles," *Archives of Environmental Health,* vol. 17 (September 1968), pp. 334–242; P. Chovin, "Carbon Monoxide: Analysis of Exhaust Gas Investigations in Paris," *Environmental Research,* vol. 1 (October 1967), pp; 198–216; and G.D. Clayton et al., "A Study of the Relationship of Street Level Carbon Monoxide Concentrations to Traffic Accidents," *American Industrial Hygiene Association Journal,* vol. 21 (February 1960),pp. 46–54.

10. P.K. Astrup et al., "Effects of Carbon Monoxide on the Arterial Walls" (Paper presented at the New York Academy of Sciences Conference on Biological Effects of Carbon Monoxide, New York City, January 12–14, 1970).

11. S.M. Ayres et al., "Carboxyhemoglobin: Hemodynamic and Respiratory Responses to Small Concentrations," *Science,* vol. 149 (3680) (July 9, 1965), pp. 193–194; and Idem, "Systemic and Myocardial Hemodynamic Responses to Relatively Small Concentrations of Carboxyhemoglobin (COHB)," *Archives of Environmental Health,* vol. 18 (April 1969), pp. 699–209.

12. E.W. Anderson et al., "Effect of Low Level Carbon Monoxide Exposure on Onset and Duration of Angina Pectoris," *Annal of Internal Medicine,* vol. 79 (1973), pp. 46–50; W.S. Aranow and M.S. Isbell, "Carbon Monoxide Effect on Exercise-Induced Angina Pectoris," *Annal of Internal Medicine,* vol. 79 (1973), pp. 391–95; and D. Bartlett, "Effects of Carbon Monoxide on Human Physiological Processes," *Proceedings of the Conference on Health Effects of Air Pollutants* (Washington, D.C., November 1973), pp. 112–14.

13. S.I. Cohen et al., "Carbon Monoxide and Myocardial Infarction," *Archives of Environmental Health,* vol. 19 (April 1969), pp. 510–17.

14. H.H. Hexter and J.R. Goldsmith, "Air Pollution and Daily Mortality," *American Journal of Medical Science,* vol. 241 (May 1961), pp. 581–88.

15. E.J. Cassel et al., "Air Pollution, Weather and Illness in Children and Adults in New York City Population," *Archives of Environmental Health,* vol. 18 (April 1969), pp. 523–30.

16. U.S. Environmental Protection Agency, *Air Quality Criteria for Nitrogen Oxides* (Washington, D.C.: U.S. Government Printing Office, January 1971), p. 9–19.

17. Ibid., pp. 9–10 to 9–14.

18. Ibid., p. 9–20.

19. Ibid., pp. 9–15 to 9–17.

20. R. Ehrlich and M.C. Henry, "Chronic Toxicity of Nitrogen Dioxide: I. Effects on Resistance to Bacterial Pneumonia," *Archives of Environmental Health*, vol. 17 (1968), pp. 860–65; and R. Ehrlich et al., "Influence of Nitrogen Dioxide on Resistance to Respiratory Infection: II. Effect of Nitrogen Dioxide," *Journal of Infectuous Disease*, vol. 113 (1963), pp. 72–75.

21. C. Shy et al., "The Chattanooga School Study: Effects of Community Exposure to Nitrogen Dioxide. Methods, Description of Pollutant Exposure and Results of Ventilatory Function Testing," *Journal of the Air Pollution Control Association*, vol. 20 (September 1970), pp. 283–88; and M.E. Pearlman et al., "Nitrogen Dioxide and Lower Respiratory Illness," *Pediatrics*, vol. 47 (February 1971) pp. 391–98.

22. C. Shy, "The Chattanooga School Study."

23. M.E. Pearlman et al., "Nitrogen Dioxide and Lower Respiratory Illness."

24. J.M. Heuss et al., "National Air Quality Standards for Automotive Pollutants—A Critical Review," *Journal of the Air Pollution Control Association*, vol. 21 (1971), pp. 535–48.

25. D.S. Barth et al., "Discussion," *Journal of the Air Pollution Control Association*, vol. 21 (1971), pp. 544–48.

26. U.S. Department of Health, Education and Welfare, *Air Quality Criteria for Photochemical Oxidants* (Washington, D.C.: U.S. Government Printing Office, March 1970), pp. 8–1 to 8–10.

27. Ibid., pp. 8–10 to 8–12.

28. Ibid., pp. 8–25 to 8–35.

29. Ibid., p. 25.

30. Ibid., pp. 8–18 to 8–25.

31. Ibid., p. 25.

32. Ibid., pp. 8–38 to 8–39 and 9–14 to 9–20, and W.S. Wayne et al., "Oxidant Air Pollution and Athletic Performance," *Journal of the American Medical Association*, vol. 199 (March 20, 1967), pp. 901–04.

33. D.I. Hammer,"Los Angeles Student Nurse Study," *Archives of Environmental Health*, vol. 28 (May 1974), pp. 255–60.

34. U.S. Department of Health, Education and Welfare, *Air Quality Criteria for Photochemical Oxidants*, pp. 9–7 to 9–12.

35. C.E. Schoettlin and E. Landau, "Air Pollution and Asthmatic Attacks in the Los Angeles Area," *Public Health Reports*, vol. 76 (1961), pp. 545–81.

36. U.S. Department of Health, Education and Welfare, *Air Quality Criteria for Photochemical Oxidants*, pp. 9–1 to 9–7; and H. Ury, "Photochemical Air Pollution and Automobile Accidents in Los Angeles," *Archives of Environmental Health*, vol. 17 (September 1968), pp. 334–42.

37. R. Hausknecht, "Air Pollution Effects Reported by California Residents," (Berkeley, California: Department of Public Health, 1960); and P. Buell et al.,"Cancer of the Lung and Los Angeles Type Air Pollution: Prospective Study," *Cancer*, vol. 10, (December 1967), pp. 2139–47.

38. P. Buell et al., "Cancer of the Lung and Los Angeles Type Air Pollution: Prospective Study."
39. U.S. Department of Health, Education and Welfare, *Air Quality Criteria for Particulate Matter* (Washington, D.C.: U.S. Government Printing Office, January 1969), pp. 51-62.
40. E. Robinson, "Effects of Air Pollution on Visibility," in A.C. Stern, ed., *Air Pollution*, 2nd ed. (New York: Academic Press, Inc., 1968), pp. 349-400.
41. U.S. Department of Health, Education and Welfare, *Air Quality Criteria for Particulate Matter*, p. 105.
42. Ibid.
43. Idem, *Air Quality Criteria for Photochemical Oxidants*, pp. 9-27 to 9-29.
44. Idem, *Air Quality Criteria for Carbon Monoxide*, p. 2-3.
45. U.S. Environmental Protection Agency, *Air Quality Criteria for Nitrogen Oxides*, p. 2-2.
46. Ibid., p. 2-5.
47. Ibid.
48. U.S. Department of Health, Education and Welfare, *Air Quality Criteria for Photochemical Oxidants*, p. 8-39.
49. Ibid., p. 8-38.
50. Ibid.
51. Ibid., p. 9-16 to 9-18.
52. Ibid., p. 9-18.
53. Idem, *Air Quality Criteria for Carbon Monoxide*, p. 2-6.
54. U.S. Environmental Protection Agency, *Air Quality Criteria for Nitrogen Oxides*, p. 7-2.
55. Ibid., p. 7-6.
56. Two other fading problems caused by NO_x have been less well researched and documented. Fading of cotton and rayon in blue and green shades has been reported from field studies and verified in laboratory studies. However, these effects only occur in high humidity conditions. In addition, yellow discoloration of undyed white or pastel colored fabrics, mostly in permanent press materials, has been reported. This discoloration is thought to result from chemical reactions of NO_x and certain of the additives, such as optical brighteners, soil release finishers, and softeners. Resistant additives can be used to avoid these discoloration effects: but again these are more costly and therefore result in higher fiber-using product prices.
57. Ibid., p. 7-2.
58. M.A. Morris et al., "The Effect of Air Pollutants on Cotton," *Textile Research Journal*, vol. 34 (June 1964), pp. 563-64.
59. U.S. Environmental Protection Agency, *Air Quality Criteria for Nitrogen Oxides*, p. 7-5.
60. Ibid., p. 7-6.
61. Ibid.
62. H.W.Hermance et al., "Relation of Air-Borne Nitrate to Telephone Equipment Damage," *Environmental Science and Technology* (1970).

63. U.S. Environmental Protection Agency, *Air Quality Criteria for Nitrogen Oxides*, p. 7–6.

64. U.S. Department of Health, Education and Welfare, *Air Quality Criteria for Photochemical Oxidants*, p. 7–1.

65. Ibid., p. 7–3.

66. The Rubber Industry, *Rubber Age*, Vol. 101 (January 1969), pp. 45–47.

67. U.S. Department of Health, Education and Welfare, *Air Quality Criteria for Photochemical Oxidants*, p. 7–2.

68. Ibid.

69. M.A. Morris et al., "The Effect of Air Pollution on Cotton," *Textile Research Journal*, vol. 34 (June 1964), pp. 563–64.

70. V.S. Salvin and R.A. Walker, "Service Fading of Disperse Dyestuffs by Chemical Agents Other Than Oxides of Nitrogen," *Textile Research Journal*, vol. 25 (July 1955), pp. 571-82.

71. U.S. Department of Health, Education and Welfare, *Air Quality Criteria for Photochemical Oxidants*, p. 7–5.

72. Idem, *Air Quality Criteria for Carbon Monoxide*, p. 7–1.

73. Ibid., pp. 7–1 to 7–3.

74. U.S. Environmental Protection Agency, *Air Quality Criteria for Nitrogen Oxides*, pp. 8–1 to 8–2.

75. Ibid., p. 8–1.

76. R.A.B. Glater, "Smog and Plant Structure in Los Angeles County," Report No. 70–17, (Los Angeles, California: University of California at Los Angeles School of Engineering and Applied Science, March 1970).

77. W.W. Heck, "Plant Injury Induced by Photochemical Reaction Products of Propylene-Nitrogen Dioxide Mixtures," *Journal of the Air Pollution Control Association*, vol. 14 (July 1964), pp. 255-61.

78. J.T. Middleton and A.O. Paulus, "The Identification and Distribution of Air Pollution through Plant Response," *Archives of Industrial Health*, vol. 14 (December 1956), pp. 526–32.

79. U.S. Department of Health, Education and Welfare, *Air Quality Criteria for Photochemical Oxidants*, p. 6–1.

80. Ibid., p. 6–2.

81. Ibid., p. 6–3.

82. Ibid.

83. Ibid., p. 6–2.

84. Ibid., p. 6–3.

85. W.A. Feder and F.J. Campbell, "Influence of Low Levels of Ozone on the Flowering of Carnations," *Phytopathology*, vol. 58 (July 1968), pp. 1038–39.

Appendix B

List of States by Region

Northeast Region

Connecticut
Maine
Massachusetts
New Hampshire
New Jersey
New York
Pennsylvania
Rhode Island
Vermont

North Central Region

Illinois
Indiana
Iowa
Kansas
Michigan
Minnesota
Missouri
Nebraska
North Dakota
Ohio
South Dakota
Wisconsin

South Region

Alabama
Arkansas
Delaware
District of Columbia
Florida
Georgia
Kentucky
Louisiana
Maryland
Mississippi
North Carolina
Oklahoma
South Carolina
Tennessee
Texas
Virginia
West Virginia

West Region

Arizona
California
Colorado
Idaho

West Region(Continued)

Montana
Nevada
New Mexico
Oregon
Utah
Washington
Wyoming

SOURCE: U.S. Bureau of the Census, *Census of Housing: 1970 Metropolitan Housing Characteristics,* Final Report HC(2)-1 (Washington, D.C.: U.S. Government Printing Office, 1972), p. xi.

References

Ad Hoc Committee on the Cumulative Regulatory Effects on the Cost of Automobile Transportation. *Final Report.* Washington, D.C.: U.S. Government Printing Office, 1972.

Austin, T.C., and Hellman, K.M. "Passenger Car Fuel Economy—Trends and Influencing Factors." Paper number 730790, Society of Automotive Engineers, Detroit, Michigan:, September 1973.

Chase Econometric Associates, Inc. *Phase II of the Economic Impacts of Meeting Exhaust Emission Standards 1971–1980.* Springfield, Virginia: National Technical Information Service, December 1971.

Grad, F.P. et al. *The Automobile and the Regulation of Its Impact on the Environment.* Norman, Oklahoma: University of Oklahoma Press, 1975.

Heywood, J.P. "Future Emission Control Technology." In Grad, F.P. et al., pp. 179–324.

Heywood, J.P. "Impact of Emission Controls: 1968–1974." In Grad, F.P. et al., pp. 115–150.

Heywood, J.P. "Inspection/Maintenance and Retrofit of In-Use Automobiles." In Grad, F.P. et al., pp. 231–278.

Ingram, G.D., and Fauth, G.R. *TASSIM: A Transportation and Air Shed Simulation Model.* Final Report to the U.S. Department of Transportation. Springfield, Virginia: National Technical Information Service, May 1974.

Internal Revenue Service. *Statistics of Income–1970, Individual Income Tax Returns.* Washington, D.C.: U.S. Government Printing Office, 1972.

Jacoby, H.D., and Steinbruner, J.W. *Clearing the Air.* Cambridge, Massachusetts: Ballinger Publishing Company, 1973.

Katona, George et al. *1970 Survey of Consumer Finances.* Ann Arbor, Michigan: University of Michigan, 1971.

Lave, L.B., and Seskin, E.P. "Air Pollution and Human Health." *Science* 169 (August 21, 1970): 723–733.

Lindgren, Leroy H. *Supplemental Report on Manufacturability and Costs of Proposed Low-Emission Automotive Engine Systems.* Washington, D.C.: National Academy of Sciences, January 1973.

Michigan Survey Research Center. *Data from the 1970 Survey of Consumer Finances.* Ann Arbor, Michigan: University of Michigan, 1973.

Motor Vehicles Manufacturers' Association. *1973/74 Automobile Facts and Figures.* Detroit, Michigan: Motor Vehicles Manufacturers' Association, 1974.

Musgrave, Richard A. et al. "The Distribution of Fiscal Burdens and Benefits." Harvard Institute of Economic Research Discussion Paper Number 319, Harvard University, Cambridge, Massachusetts, September 1973.

National Academy of Sciences. *Proceedings of the Conference On Health Effects of Air Pollutants.* Prepared for the Committee on Public Works, U.S. Senate. Washington, D.C.: U.S. Government Printing Office, November 1973.

National Academy of Sciences. *A Report by the Committee on Motor Vehicle Emissions.* Washington, D.C.: National Academy of Sciences, February 12, 1973.

National Academy of Sciences. *A Report by the Coordinating Committee on Air Quality Studies.* Prepared for the Committee on Public Works, U.S. Senate. *Volume IV: The Costs and Benefits of Automobile Emission Control.* Washington, D.C.: U.S. Government Printing Office, September 1974.

Ohta, Makoto, and Griliches, Zvi. "Automobile Prices Revisited: Extentions of the Hedonic Hypothesis." Harvard Institute of Economic Research Discussion Paper Number 325, Harvard University, Cambridge, Massachusetts, October 1973.

Ridker, R.G., and Henning, J. "The Determinants of Residential Property Value with Special Reference to Air Pollution," *Review of Economics and Statistics* 49 (May 1967): 246–257.

U.S. Bureau of the Census. *Census of Housing: 1960. Metropolitan Housing.* Final Report HC(2)-1 to HC(2)-200. Washington, D.C.: U.S. Government Printing Office, 1962.

U.S. Bureau of the Census. *Census of Housing: 1970 Metropolitan Housing Characteristics.* Final Report HC(2)-1 to HC(2)-244. Washington, D.C.: U.S. Government Printing Office, 1972.

U.S. Council of Environmental Quality. *Environmental Quality: The Third Annual Report.* Washington, D.C.: U.S. Government Printing Office, 1972.

U.S. Department of Health, Education and Welfare, *Air Quality Criteria for Carbon Monoxide.* Washington, D.C.: U.S. Government Printing Office, March 1970.

U.S. Department of Health, Education and Welfare. *Air Quality Criteria for Hydrocarbons.* Washington, D.C.: U.S. Government Printing Office, March 1970.

U.S. Department of Health, Education and Welfare. *Air Quality Criteria for Particulate Matter.* Washington, D.C.: U.S. Government Printing Office, January 1969.

U.S. Department of Health, Education and Welfare. *Air Quality Criteria for Photochemical Oxidants.* Washington, D.C.: U.S. Government Printing Office, March 1970.

U.S. Department of Health, Education and Welfare. *Control Techniques for Carbon Monoxide, Nitrogen Oxide, and Hydrocarbon Emissions from Mobile Sources.* Washington, D.C.: U.S. Government Printing Office, March 1970.

U.S. Department of Transportation. *Annual Miles of Automobile Travel.* Report No. 2 of Nationwide Personal Transportation Study. Washington, D.C.: U.S. Department of Transportation, April 1972.

U.S. Department of Transportation. *Household Travel in the United States,* Report No. 7 of the Nationwide Personal Transportation Study. Washington, D.C.: U.S. Department of Transportation, December 1972.

U.S. Environmental Protection Agency. *Air Quality Criteria for Nitrogen Oxides.* Washington, D.C.: U.S. Government Printing Office, January 1971.

U.S. Environmental Protection Agency. *Annual Report of the Administrator of the Environmental Protection Agency.* Washington, D.C.: U.S. Government Printing Office, 1973.

U.S. Environmental Protection Agency. *Modified Rollback Comptuer Program Documentation.* Washington, D.C.: Environmental Protection Agency, November 1973.

U.S. Environmental Protection Agency. *The National Air Monitoring Program: Air Quality and Emissions Trends.* Annual Report, vol. 1. Research Triangle Park, North Carolina: U.S. Environmental Protection Agency, August 1973.

U.S. Environmental Protection Agency. *Supplement Number 2 for Compilation of Air Pollutant Emission Factors.* Report AP-42, 2nd ed. Washington, D.C.: U.S. Government Printing Office, 1973.

Wykoff, F.C. "A User Cost Approach to New Automobile Purchases." *Review of Economic Studies XL* (July 1973): 377–390.

Index

air pollutants: aesthetic effects, 148; benefits analysis, 95–100; damages, 138; dollar benefit, 137; health effects, 141; health issues, 140; multicar strategy, 113; plant effects, 151; stringent strategy, 133; in urban areas, 19; urban emission rates, 69, 71; usage analysis, 24–27

air quality: benefits, 96; willingness to pay, 138

auto companies: compliance, 36; emission control cost transfer, 12; fuel economy penalties, 48; lobby effort, 2; new car prices, 40; profits, stockholders, and taxes, 59–61; profit strategy, 82; stockholder costs, 16

Baltimore: relative cost burden, 87

benefits: air quality calculation, 20; assessment, 95; comparative for income groups, 22; cost/benefit quality improvement, 17; distribution and propoor ratio, 105; estimation, 18; geographic analysis, 72; propoor ratio, 23; strategy analysis, 123; urban areas, 103

Birmingham, 83, 84

Boston, 83, 84; CO benefits, 101; relative cost burden, 87

burden, concept of, 21

California: effort in Los Angeles, 1; health effects of pollutants, 147, 149

car ownership: burdens, 59; car age and income group, 24; car age and operating costs, 15; cost analysis, 53; costs by car age, 14; fuel economy penalties, 48; maintenance, 45; new car prices, 41; operating costs, 56; travel mileage and income group, 27; used cars, 43

Cassell, E.J. et al., 145

catalytic converter, 37; in Heywood, 47

central city: air quality benefits and variations, 100; CO benefit, 102; cost pattern, 83; emission rates, 19, 68; models, 10; ownership probabilities, 25; two car scheme, 110–112; urban modeling procedure, 11; variations by population, 130

Chase Econometric Associates, 41

Chattanooga: NO_2, 146

Chovin, P., 143

Cincinnati: NO_x effects, 150

Clayton, G.D. et al., 143

Clean Air Act, 1963, 1970 Amendments, 1; research review, 9; time periods, 36

Cleveland: NO_x effects, 150

CMVE (Committee on Motor Vehicle Emission), 34; projections, 46–49

CO (carbon monoxide): 1978, 10; aesthetic effects, 149; city variations, 130; $COHb_x$ damages, 142; concentration from cars, 17; concentration by city, 104; damages, 141; emission rate analysis, 21, 63, 64; federal reduction expenditure, 127; new car rate, 66–68; reduction, 36, 95

Cohen, S.I., et al., 144

costs: alternative strategies, 115; burden and population size, 86; car operating, 15; categories, 13; concept of for emission control, 9; concept by relative, 21; dual catalyst system, 38; estimation strategy, 10; distribution, 129; housing and property values, 140; regressive cost burdens, 128; used cars, 43; windfall gains, 45

About the Author

David Harrison, Jr. is Assistant Professor of City and Regional Planning at Harvard University. He received the B.A. in economics from Harvard University, the M.Sc. in the economics of transport from the London School of Economics, and the M.A. and Ph.D. in economics from Harvard University. He is a co-author of *The Automobile and the Regulation of Its Impact on the Environment* (Norman, Oklahoma: University of Oklahoma Press, 1975) and contributed to the recent National Academy of Sciences Study, *The Costs and Benefits of Automobile Emission Control* (Washington, D.C.: U.S. Government Printing Office, 1974). He has also written articles for professional journals in the general field of urban affairs.